DATE DUE			
May 22 78			
Mar 27 79			

ANALYTICAL METHODS APPLIED TO
AIR POLLUTION MEASUREMENTS

ROBERT K. STEVENS
Research Chemist

and

WILLIAM F. HERGET
Research Physicist

Chemistry and Physics Laboratory
National Environmental Research Center
U.S. Environmental Protection Agency
Research Triangle Park, North Carolina

 ann arbor science PUBLISHERS INC.
POST OFFICE BOX 1425 • ANN ARBOR, MICH. 48106

Copyright © 1974 by Ann Arbor Science Publishers, Inc.
P.O. Box 1425, Ann Arbor, Michigan   48106

Library of Congress Catalog Card No. 74-77400
ISBN   0-250-40046-4

PREFACE

This volume brings together the leading scientists pioneering in the development of improved techniques to measure the properties of particulate and gaseous air pollutant species.

Their work represents the latest technological advances in chemistry, physics, and electrical engineering in the development of instrumentation to characterize undesirable components in the atmosphere. These include the application of lasers and other electro-optical devices to measure pollutant concentrations across both open paths in the atmosphere and power plant smoke stacks *without physical sampling of the pollutant*.

The broad range of subjects includes thorough discussions of the rapidly advancing technique of gas-filter correlation spectroscopy, comparison of classical with newly developed X-ray fluorescent methods applied to the measurement of elemental composition of particles, and advances in techniques for extracting gaseous and particulate pollutants from power plant stacks. The topics further include the applications of chemiluminescent techniques to measure ambient concentrations of gaseous pollutants, and descriptions of two new techniques to measure sulfuric acid vapor at ambient concentrations.

Fourteen chapters are organized in three sections: Section I deals with methods to measure gaseous pollutants at ambient concentrations, Section II relates to procedures to characterize the properties of particulates in the atmosphere, and Section III describes instrumental methods for analysis of gaseous and particulate pollutants at source concentrations.

The total book is introduced by Dr. A. P. Altshuller, who also introduced the symposium at the 165th National Meeting of the American Chemical Society in Dallas, Texas, April 8-13, 1973, from which this book emanated. We are fortunate to have his plenary lecture here, as it outlines the Environmental Protection Agency's requirements for air pollution monitoring in the 1970s and suggests various avenues that may be explored to meet these stringent requirements. This, in fact, sets the tone for the book.

It is felt that the reader will find this text useful as a review of the latest developments in air pollution measurement techniques, and that perhaps he may be inspired toward the development of other improved physical methods for characterizing the condition of our atmosphere.

The editors wish to express their appreciation to all who prepared manuscripts for this volume.

July 1974   Robert K. Stevens
                                            William F. Herget

CONTENTS

SECTION I
TECHNIQUES TO MEASURE GASEOUS POLLUTANTS AT AMBIENT CONCENTRATIONS

SECTION II

TECHNIQUES TO MEASURE CHEMICAL AND PHYSICAL PROPERTIES OF PARTICLES IN THE ATMOSPHERE

SECTION III

TECHNIQUES TO MEASURE POLLUTANTS FROM STATIONARY AND MOBILE SOURCES

LIST OF CONTRIBUTORS

A. P. Altshuller: Chemistry and Physics Laboratory, National Environmental Research Center, U.S. Environmental Protection Agency, Research Triangle Park, North Carolina.

H. M. Barnes: Chemistry and Physics Laboratory, National Environmental Research Center, U.S. Environmental Protection Agency, Research Triangle Park, North Carolina.

D. E. Burch: Aeronutronic Division, Philco-Ford Corporation, Newport Beach, California.

A. R. Calawa: Lincoln Laboratory, Massachusetts Institute of Technology, Lexington, Massachusetts.

R. L. Chuan: CELESCO Industries, Inc., Costa Mesa, California.

J. W. Davis: Department of Chemical Engineering, Pennsylvania State University, University Park, Pennsylvania.

J. N. Driscoll: Environmental Consultant, 35-7 Roughan, Revere, Massachusetts.

T. G. Dzubay: Chemistry and Physics Laboratory, National Environmental Research Center, U.S. Environmental Protection Agency, Research Triangle Park, North Carolina.

L. J. Forney: Division of Engineering and Applied Physics, Harvard University, Cambridge, Massachusetts.

F. S. Goulding: Lawrence Berkeley Laboratory, University of California, Berkeley, California.

D. A. Gryvnak: Aeronutronic Division, Philco-Ford Corporation, Newport Beach, California.

W. F. Herget: Chemistry and Physics Laboratory, National Environmental Research Center, U.S. Environmental Protection Agency, Research Triangle Park, North Carolina.

E. D. Hinkley: Lincoln Laboratory, Massachusetts Institute of Technology, Lexington, Massachusetts.

J. A. Hodgeson: Chemistry and Physics Laboratory, National Environmental Research Center, U.S. Environmental Protection Agency, Research Triangle Park, North Carolina.

J. B. Homolya: Chemistry and Physics Laboratory, National Environmental Research Center, U.S. Environmental Protection Agency, Research Triangle Park, North Carolina.

J. M. Jaklevic: Lawrence Berkeley Laboratory, University of California, Berkeley, California.

B. V. Jarrett: Lawrence Berkeley Laboratory, University of California, Berkeley, California.

H. C. Lord: Environmental Data Corporation, Monrovia, California.

W. A. McClenny: Chemistry and Physics Laboratory, National Environmental Research Center, U.S. Environmental Protection Agency, Research Triangle Park, North Carolina.

J. D. Meng: Lawrence Berkeley Laboratory, University of California, Berkeley, California.

P. S. Mudgett: Cabot Corporation, Billerica Research Center, Billerica, Massachusetts.

L. W. Richards: Cabot Corporation, Billerica Research Center, Billerica, Massachusetts.

J. R. Roehrig: Cabot Corporation, Billerica Research Center, Billerica, Massachusetts.

R. Rollins: Chemistry and Physics Laboratory, National Environmental Research Center, U.S. Environmental Protection Agency, Research Triangle Park, North Carolina.

H. C. Rook: Chemistry and Physics Laboratory, National Environmental Research Center, U.S. Environmental Protection Agency, Research Triangle Park, North Carolina.

R. K. Stevens: Chemistry and Physics Laboratory, National Environmental Research Center, U.S. Environmental Protection Agency, Research Triangle Park, North Carolina.

R. Villalobos: Beckman Instruments, Inc., Process Instruments Division, Fullerton, California.

INTRODUCTION

INSTRUMENTS FOR AIR POLLUTION MONITORING
DURING THE 1970'S

A. P. Altshuller

INTRODUCTION

Progress in the development of new instruments for air pollution measurements has been much improved during recent years. I believe this progress can be attributed in good part to the resources made available by the Air Quality Act of 1970 to the Environmental Protection Agency.

The program has carried R/D from sensor development through field evaluation of prototype instruments. We have worked closely with instrument companies to encourage rapid commercialization and marketing of prototype instruments that performed satisfactorily. As a result, flame photometric, gas chromatographic and chemiluminescent analyzers have become available for sulfur dioxide, hydrogen sulfide, organic sulfur compounds, hydrocarbons, methane, carbon monoxide, ozone, nitric oxide and nitrogen dioxide. Research is in progress on instruments for particle mass, size and composition.

The obvious difficulty in attempting predictions is the inability to see beyond the horizon. Occasionally a method with little if any prior history of utility will burst forth as a highly versatile instrumental technique. A more satisfactory approach is to concentrate on future needs rather than attempt to estimate the details of progress on individual analytical techniques. Therefore, this chapter will concentrate mainly on research needs at present and in the immediate future. Several major areas have

3

been selected for discussion. No comprehensive
listing of needs for specific instruments will be
attempted.

PORTABLE AIR QUALITY DOSIMETERS

Community health studies, to be of the greatest
effectiveness, should be accompanied by air quality
measurement capabilities optimized to meet specific
research needs. Past community health studies often
have had to use the monitoring results from networks
set up for other purposes. More recently monitoring
sites have been set up to meet the requirement of
specific health studies (CHESS).

Participants in health studies actually are exposed to a number of air quality environments, only
one of which is the community atmosphere ordinarily
monitored. The community air quality monitoring
networks represent a number of compromises. Although
a number of pollutants ordinarily are measured at
each site, it does not follow that a site selected
as appropriate for monitoring of carbon monoxide or
sulfur dioxide can be equally appropriate for monitoring ozone or nitrogen dioxide. The practical limitation on number of sites often makes it difficult or
impossible to construct concentration isopleths with
sufficiently high resolution to satisfy particular
needs. All of these problems are compounded if a
monitoring network is expected to provide concurrently
measurements for emergency episodes, trends in pollutant concentrations, health studies or transport
and transformation of pollutants research. These
difficulties again suggest the need to develop
measurement techniques specifically tailored to
health research requirements.

In health studies the integrated dosages to which
participants are exposed are of great importance.
These dosages are accumulated when individual participants are exposed not only to community atmospheres
but also to the atmospheres in their homes, work and
recreational environments and in movement between
these various environments. The most practical
approach to meeting these requirements is the
development of personal air quality dosimeters or
personal monitors.

Such dosimeters might be fountain pen or similar
sized instruments carried on the person of individual
participants. Such equipment is not presently available. Personal monitors the size and weight of

portable radios or attache cases are much closer to
the present state-of-the-art in air quality instrumentation. A number of principles already successfully developed appear directly applicable to
monitoring instruments in this size and weight range.
Such monitors could be easily operated on line or
battery power at a convenient location within home,
office, shop, or motor vehicle. Two such analyzers
might be provided per participant to provide for
maintenance, calibration and other requirements.

Measurements for emergency episodes or transport
research studies require rapid data handling justi-
fying real time analyzers and signal processing.
Health studies often do not require the same feed-back
times. A simple magnetic tape system that permits
mailing or periodic pickup might be satisfactory.
Calibration systems can be built into analyzers in
conventional monitoring sites. Unless such calibration systems can be readily miniaturized, an acceptable alternative could be periodic calibration at
a central laboratory or mobile laboratory facility.
Integrated measurement techniques with 1- to 24-hour
integration times should be acceptable. A prime
objective should be providing workable instruments
in the shortest time possible so they can be utilized
as soon as possible in community health studies.

Because individual health studies often are
concerned with the impact of one or a few pollutants,
multipollutant analyzer capability probably is not
as pertinent as in other applications. Pollutants
having high priority are sulfur dioxide, nitrogen
dioxide and ozone. Measurement of particulate
species probably is more practically approached by
personal mini-samplers with subsequent laboratory
analysis when 24-hour average or longer integration
intervals are appropriate.

The above discussion is not meant to offer
definitive criteria or guidelines for personal air
quality monitors but rather to stimulate consideration of the best approaches to meeting these
specialized needs.

PARTICLE MASS, SIZE AND COMPOSITION

Although a variety of real time air quality and
emission measurement instruments exist, very few real
time or even 1- to 2-hour average monitoring instruments exist for particles. About the only instrument
now being placed in monitoring networks that responds

to a property of particles is the integrating
nephelometer for measurement of visibility. Ordinarily sampling is done on a 24-hour average basis,
with weighing to determine mass or laboratory
analysis to determine chemical composition. By
using x-ray fluorescence techniques 2-hour average
samples often can be analyzed satisfactorily.
Particle sizing can be done with various impactors
requiring subsequent weighing and chemical analysis
of the particles collected on the several stages.
Electrical mobility and optical particle sizing
equipment are available for sizing by particle
number on a real time basis, but only research type
instruments are available (Whitby Aerosol Analyzer,
Rayco optical single particle counter) and these
instruments plus an impactor are necessary to cover
the 0.01-10 μg range.

Such research real time particle size analyzers
do not provide particle mass directly nor permit
chemical analyses. Therefore a group of various
particle size analyzers, impactors, and filters must
be used to accumulate all of the data needed for
particle mass, size and composition in comprehensive
field studies. In terms of monitoring needs for a
control program the usual request is to simplify the
measurements to one or two techniques that are simple
and inexpensive in field use. Unfortunately such
requests can be inconsistent with scientific requirements. Particles as they are emitted from sources
and as they reside in the atmosphere undergo complex
chemical and physical transformations. Gases emitted
from sources are transformed to particle species in
the atmosphere. Under these circumstances, the
ability to relate air quality for particles to source
emission control strategies can be resolved only by
going progressively from the complex to the less
complex measurement approach.

What I am suggesting is that comprehensive
measurement programs are needed in order to determine
exactly where simplifications can justifiably be
made. Furthermore, even with these simplifications
I believe it unlikely that the degree of simplifica-
tion desired by control officials can be achieved.
Whether this prediction is correct or not, it is
important that the measurement techniques be developed
and utilized as rapidly as possible to determine on
the basis of sound experimental evidence the optimum
approach to improved air quality of particulate
pollutants.

The present experimental results indicate that
a substantial fraction of the particle volume and

mass occurs in the particle size range around and
below 1 μm. The particles in this range result from
condensation of photochemically formed reaction
products on very small nuclei followed by coagulation
of very small particles particularly onto preexisting
particles in the 0.1-1 μm range. Other small particles such as lead and bromine aerosols from
vehicular exhaust also coagulate to form particles
in this range. Particles larger than 1 or 2 μm
appear to arise from primary emission sources or
fugitive dusts. Silicates, aluminum, magnesium and
much of the iron, and certain other metals in various
forms are found in the particles above 1 or 2 μm.
Certain other elements appear to be distributed more
uniformly through the particle size ranges.

The significance of particle size has several
aspects. Visibility reduction as measured by the
integrating nephelometer and other techniques are
closely related to particles between 0.1 and 1 mμ.
Laboratory studies on inhalation toxicity of sulfates
indicate an increased effect in the range at or below
1 mμ. However, the respirable range is often taken
at 5 mμ and below. The available size results for
urban aerosols characteristic now indicate a maximum
bimodal distribution with a maximum in the volume
distribution below 1 mμ and a minimum between 1 and
2 mμ.

Most of the aerosol measured below 1 mμ appears
not to have originated as aerosol but to have formed
by atmospheric reactions converting sulfur dioxide
to sulfate, nitrogen dioxide to nitrate and organic
vapors to organic aerosols. These species predominate
over lead aerosols and other primary aerosol emissions
in the <1 mμ range. If so, it is essential to be
able to associate quantitatively the air quality
composition of these aerosol species with the control
of emissions of the gaseous pollutant from which
they are formed.

Since the control must be at the source and must
be specific to chemical species, *i.e.*, sulfur dioxide,
hydrocarbons, or lead, the relationships between
degree of control and manifestation seem essential.
For example, how much must atmospheric sulfates be
reduced to achieve significant improvement in
visibility through what degree of control of sulfur
dioxide at emission sources?

It seems necessary to be able to measure mass in
at least two preselected size ranges and to be able
to determine chemical composition of the particles
in these size ranges. Concurrently, visibility
should be correlated with size and composition.

Since aerosol volume appears to be a function of
relative humidity based on laboratory and field
observations, a significant fraction of the mass of
submicron liquids may be associated with condensed
water at higher relative humidities. Other volatile
species may be associated with the submicron particles. Volatile constituents can be lost on collection in filters on impactor surfaces. Therefore,
in situ particle measurements may prove to give
higher levels of submicron particles than do methods
based on collection on surfaces.

At the present time organic aerosols are particularly poorly characterized. The only routine
measurement made is of benzene-soluble material from
glass fiber filters. In addition to possible losses
of volatiles off filter surfaces, benzene will extract only a part of the organic aerosols. Polar
organic aerosol materials are inefficiently extracted
by benzene. Therefore, it appears that the available
results significantly underestimate the mass of
atmospheric organic aerosols. Another aspect of the
problem associated with organic aerosols is the
differing abilities of various organic vapors to
participate in aerosol formation.

Work is underway to develop methods for determining sulfuric acid, sulfate and nitrate. Total
sulfur can be determined by x-ray techniques, as can
lead and some other species present in the submicron
range. Since continuous gas analyzers are accepted
as essential for gaseous pollutants, it would appear
that continuous or semicontinuous analyzers for
sulfur, nitrogen and organic particulates will come
into demand. At present the emphasis continues to
be directed toward improving the reliability of
sampling in the field, with much of the analysis
conducted at a central laboratory.

For particles with metastable characteristics,
in situ sizing and chemical analysis would be ideal.
In situ chemical analysis of individual particles
should offer a real challenge to our analytical and
instrumental ingenuity.

ELECTROOPTICAL MEASUREMENTS

Application of electrooptical techniques to
measurement of air pollution is attractive for several
reasons: (1) it avoids use of liquid or gaseous
reagents, (2) it provides the potential for multipollutant analysis, (3) it provides *in situ* capabiliti

for source emission analysis, and (4) it provides
long-path and remote measurement capabilities. Each
of these aspects will be considered separately. The
first advantage is self-evident and it has stimulated
past and present developments of infrared nondispersive instruments for carbon monoxide and carbon
dioxide and ultraviolet analyzers for sulfur dioxide
and nitrogen dioxide. Improved approaches to nondispersive analysis and use of infrared and ultraviolet fluorescence and laser sources all provide
impetus to continued research activity. A particularly urgent need is for an electrooptical nitrogen
dioxide analyzer for atmospheric analysis applications.
    Multipollutant capability in principle is highly
attractive. In practice analytical techniques, whether
electrooptical or other, have proved to be more
applicable to one or a few pollutants. There are
several practical considerations that need to be
stated. An important area of application—air quality
monitoring—will be used as an example. If the
application is to be air quality measurement to meet
the air quality standards, there are five gaseous
pollutants susceptible to measurement: carbon
monoxide, sulfur dioxide, nitrogen dioxide and
methane. Total particulates represent a separate
analytical problem. Although electrooptical techniques can be used to measure certain size ranges of
particulate matter, it appears that a separate
analyzer would be required whatever the analytical
approach. Total hydrocarbon measurement as a reference method and as a class analysis is now defined
in terms of flame ionization analysis for ambient
air and for source emissions. It is not clear that
any electrooptical measurement could give equivalent
results.
    Putting aside these particular difficulties the
following type of questions arise consistently in
comparing multipollutant analyzers: (1) Will the
multicomponent analyzer have sensitivity, specificity,
linear range, stability, response time compatible with
the requirements for each of the five components?
(2) How do the capitalization costs per component
for the multipollutant analyzer compare with the
corresponding alternative single pollutant analyzers?
(3) What will be the maintenance requirements and
downtime statistics? If a single analyzer replaces
five analyzers, it must have minimal downtime to
avoid total loss of all outputs. (4) What level of
technical expertise is necessary to insure continuing
operation aside from major malfunctions?

A practical, marketing type aspect involved for
any analyzer, but particularly for a high cost multipollutant electrooptical instrument, is the prior
capitalization situation. If agencies already have
invested substantial sums in an array of single
pollutant analyzers, they are not in a position for
a number of years to justify replacement with a new
system even if it meets all of the above requirements
under field conditions. Even if one or two single
pollutant analyzers are less than completely satisfactory, substitution with better single pollutant
analyzers tends to be the usual decision rather than
complete recapitalization. For these reasons it
appears that even an outstanding new multipollutant
analyzer system could be slow in gaining usage compar
to new single or dual pollutant analyzers.

In stationary source emission measurement, *in sit*
stack electrooptical instruments are practical.
Several such instruments already exist for measuring
sulfur dioxide, nitrogen oxides and opacity. The
alternative is the use of air quality analyzers.
However, these analyzers even with modification tend
to be difficult to operate on an untreated gas stream
from an emission source. High particulate loadings,
high water contents, high temperatures, and corrosive
substances alone and in combination tend to make many
air quality analyzers inoperative. Therefore, the
original gas stream usually needs to be cooled,
filtered and most of the water removed. The result
is a relatively complex sampling interface between
source and analyzer. Since the *in situ* analyzers
must be specifically designed for the "hostile"
environment in the stack, they must be tailored to
fit the specific applications.

Open path and remote uses of electrooptical
techniques constitute an active field of research.
Experience indicates that the development and
evaluation of such techniques can be very expensive
and time-consuming. Therefore, the justifications
for development must be based on much more than
scientific interest or enthusiasm to develop a new
instrument. Valid justifications would include the
following:

1. rapid determination of compliance of stack
    emissions or "extended" industrial emissions
    source with standards from the fence-line or
    by aircraft measurements

2. obtaining three-dimensional profiles of a few
    critical pollutants to assist in verification

of air quality simulation models. Such models predict the average concentration with time in air pockets with dimensions from 100 meters to several thousand meters. Verification by convenient point-source analyzers is complicated by the fact that the concentrations predicted by a model are integrated values for a volume element usually not consistent with point sampling.

3. Verification that certain air quality measurements are correct representations of the atmospheric composition—for example, determining whether an integrating nephelometer as a point source instrument properly represents visibility in any particular area. Of course, in this particular example there is the added difficulty of evaluating the subjective impressions or desires of the community about adequate visibility.

It is important not to confuse good intentions with the ability to obtain useful results. For example, remote measurements from earth resources satellites are proving useful in a number of applications. However, air quality composition varies significantly through each day and from day to day. Either a continuous measurement of air quality is needed or an integrated measurement over an appropriate time interval. Furthermore, the air mass of interest is the very thin innermost layer next to the earth's surface. Measurement of the air quality of this layer through the earth's atmosphere is a difficult if not impossible task. More appropriate although less critical applications would be to geophysical measurements of the average concentrations in large volumes of air over oceans, plains, polar regions, etc. This type of application also minimizes the cloud cover restrictions, which are significant over particular urban areas.

SAMPLING PROCEDURES

No analysis is any better than the sample obtained. This truism most certainly is applicable to air pollution measurements. Sampling difficulties more often than not can be the actual causes of analytical anomalies. Within the probe and sampling lines gases can be adsorbed and particles impact on surfaces, and highly reactive pollutants can consume each other. These problems exist whether there is a sensor, impinger, filter or impactor surface at the end of

the measurement system. Advantages presumably gained
by use of advanced instrumentation can be negated
by flaws in sampling. There are additional problems
whenever filter surfaces are used to collect particulates prior to chemical analysis. Filter or surface
composition, impurities in or on the surfaces, loss
of volatile liquid aerosols, or oxidation of labile
species can seriously compromise the analytical
performance. In adapting analytical methods to
different emission sources, maintaining analytical
reliability can require tailoring of the sampling
interface to individual sources.

While this sort of problem can be eliminated by
the use of the *in situ* or open path electrooptical
instruments, these instruments also have associated
problems. Therefore, it is unlikely that the "extractive" techniques requiring sampling will be
replaced entirely in the foreseeable future.
Therefore, improvements in sampling procedures must
continue to occur concurrently with research on
improved instrumental techniques.

CONCLUSION

The topics discussed represent important areas
for future R/D, but they certainly are not meant to
represent the complete list of possibilities. I
deliberately have avoided discussion of areas presently receiving a great deal of attention that may
be rather well-exploited in the near future. R/D
programs, as all other programs in regulatory
agencies, are strongly impacted by legislative
mandates. These mandates often afford minimal time
to respond. When standards must be developed rapidly,
all technical activities are hard pressed to respond.
The current state of the air in measurement as in
other technical areas must be utilized. Promising
new methods that have not received adequate field
evaluation can lose out in such situations. Therefore, a large number of good, but incomplete pieces
of research can end up having no impact on standards.
It is better to make the difficult decisions early
and move forward rapidly on a smaller selection of
procedures. These sorts of considerations require
hard evaluations of lead time as well as costs on
instruments no matter how attractive they may be
scientifically. Despite these restrictions, I
believe that most promising analytical techniques
offered for consideration have been evaluated.

In concluding it must be emphasized that quality assurance and well-trained analysts or instrument personnel are critical aspects of the overall problem. While improved instruments and automated calibration systems limit the amount of human intervention, well-trained staff will continue to be essential to successfully conduct measurement and monitoring programs.

SECTION I

TECHNIQUES TO MEASURE GASEOUS POLLUTANTS
AT AMBIENT CONCENTRATIONS

CHAPTER 1

ROLE OF GAS CHROMATOGRAPHY
IN AIR POLLUTION MONITORING

R. Villalobos

INTRODUCTION AND BACKGROUND

Some four or five years ago, the Air Pollution Control Administration, anticipating the need for more sensitive air pollution measurements, developed criteria for a gas chromatographic analyzer for atmospheric methane and CO.[1,2] The need for these measurements grew out of our understanding of the role of reactive hydrocarbons in the formation of photochemical smog. The realization that methane is a nonreactive hydrocarbon and is present worldwide at background levels in excess of 1 ppm underlined the need for a sensitive method for measuring the nonmethane fraction of atmospheric hydrocarbon pollutants. Foreseeing the wide deployment of unattended ambient air monitoring stations equipped with continuous, air monitoring analyzers, EPA stimulated manufacturers of analytical instrumentation to develop a fully automated gas chromatograph suitable for such service.

It was the process analyzer manufacturers, primarily, who responded. The extensive technology that had been accumulated in providing automated chromatographs for the process industries was applied to this challenge. In the last two years fully automated gas chromatographs for continuous monitoring of ambient air hydrocarbon pollutants have appeared on the market. These chromatographs have been specifically designed for continuous, unattended operation in ambient air monitoring stations and, as such, are completely self-contained, requiring only power and a supply of gases.

Since their introduction, a number of these chromatographs have been acquired by federal, state, and local agencies for evaluating and testing.   An additional large number have been acquired for monitoring the ambient air on a continuous basis, and many of these are already in service.

As anticipated, the intrinsic value and potential usefulness of a fully automated gas chromatograph has become apparent to workers in many other fields. Apart from the numbers that have been acquired for the purpose for which it was designed, a growing number are being modified for other specialized fields of study.

The purpose of this paper is twofold:

1. to review the state-of-the-art and recent improvements in the ambient carbon monoxide/ nonmethane-hydrocarbon monitoring gas chromatograph
2. to describe a number of the other applications in which this chromatograph has been modified and applied.

Within those narrow limits, this paper will attempt to provide information of a general nature and to stimulate further activity along those and similar lines of endeavor.

It is not a purpose of this paper to review the use of gas chromatography, generally, in air pollution and related studies.   The application has been so extensive, and its value so well established and recognized, that no further stimulus is needed.

AMBIENT CO/NONMETHANE HYDROCARBON
MONITORING GAS CHROMATOGRAPH

EPA has established national primary and secondary air standards that limit CO to 9 ppm for eight hours and 35 ppm for one hour, not to be exceeded more than once per year.   Similarly, nonmethane hydrocarbons are limited to 0.24 ppm for three hours concentration (6 a.m. to 9 a.m.), not to be exceeded more than once per year.   Since the measurement for nonmethane hydrocarbons cannot be made directly, the background methane must be separately determined and subtracted from a total hydrocarbon measurement.

Basic techniques embodied in these measurements have been described elsewhere.[3-7]   Conversion of carbon monoxide to methane over a catalyst for de- tection by a flame ionization detector was used to

measure CO in urban air samples.[6],[7]   A prototype
automated chromatograph was later developed at NAPCA.[1],[2]
That instrument also measured methane, but not total
hydrocarbons. The total hydrocarbon measurement was
later developed by NAPCA: a separate sample valve
to inject a sample of ambient air directly into the
flame ionization detector (FID) to give a single
measurement indicative of the total hydrocarbons in
the air.

A modified method was developed by Villalobos
and Chapman[8] using hydrogen carrier gas for the
separation of methane and carbon monoxide, and to
supply the methanator. A separate total hydrocarbon
channel used air that had passed through a catalytic
oxidizer to remove residual hydrocarbons including
methane, and to give a "zero hydrocarbon" reference
against which the total hydrocarbon content of the
sample was measured. Provision was also included
for measuring acetylene and ethylene by means of a
third column channel. This method was used in commercially available chromatographs by a number of
manufacturers in the United States. Details regarding its design and performance have been published
elsewhere.[9]

Recent Improvements

Since its introduction, a number of changes have
been made to improve performance and overcome some
problems encountered in the field. A number of these
changes were directed at improving the stability of
the Molecular Sieve columns used for the separation
of CO and methane. Early field experience indicated
slow but progressive contamination and deactivation
of that column, resulting in unstable elution times.
This required frequent calibration and adjustments
to the component gates. Sources of contamination
were found to be:

1. deterioration of the stripper column (pre-
   cutter), which is used to separate and back-
   flush the C_2-and-heavier hydrocarbons,
   moisture and CO_2, and prevent them from
   entering the molecular sieve columns.
   Repeated contact of the material (Porapak)
   with the oxygen in the sample apparently
   resulted in breakdown of the porous polymer
   into more volatile fragments that were
   carried into the sieve column and deactivated
   it.

2. diffusion of atmospheric water into the
   hydrogen carrier gas through the diaphragm
   of the hydrogen carrier gas regulator.
   This added sufficient water to the carrier
   gas to deactivate the molecular sieve.

The first of these problems was overcome by re-
placing the Porapak column with a combination column,
which consists of a section of silica gel preceded
by a section of partition column.   The latter uses
an inert Teflon solid support, and its function is
to retain the water and prevent its entering the
silica gel column.   The silica gel column retains
CO_2 and the C_2-and-heavier hydrocarbons.   The second
problem was overcome by using a carrier gas regulator
with a nonpermeable stainless steel diaphragm.
Another problem was the close separation between
methane and the preceding air peak.   A desirable
separation is one providing a sufficient flat base
line immediately preceding the methane peak to permit
proper functioning of the "auto zero" circuit.   This
circuit, activated by the programmer, rezeroes the
base line immediately prior to the elution of the
methane peak.   An improvement in this separation was
achieved by means of a specialized activation procedure for the molecular sieve column.   The separation
was improved to the required degree and appears on
instruments currently being supplied as shown in
Figure 1.1.   An added benefit of this improved
separation has been that a larger sample can now be
used to attain more sensitivity.   Instruments are
currently being supplied with a capability of less-
than-1 ppm full scale for CO; before this improvement
even the 1 ppm was difficult to achieve.
As described in References 8 and 9, the capability
of measuring ethylene and acetylene can be added by
means of an optional third column system.   It was
originally intended to provide the additional
measurements within the five-minute cycle, but it
was found that a longer time cycle was desirable to
avoid reduced stability.   Current instruments with
this additional capability are now being supplied
with a ten-minute cycle, as shown in Figure 1.2.

Supply Gases

A number of problems encountered in early field
experience were caused by impurities in the gases
and/or lack of cleanliness of regulator and external

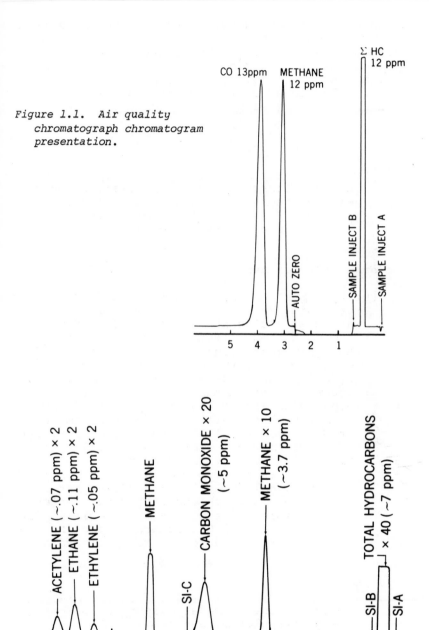

Figure 1.1. Air quality chromatograph chromatogram presentation.

Figure 1.2. Air quality chromatograph 5-component analysis.

connecting tubing. Notwithstanding the widespread
use of gas chromatographs with flame ionization
detectors, awareness of the purity of gases and
cleanliness of regulators and of the extreme installation care required for proper performance is
not equally widespread. Current recommendations
relating to these supply gases follow.

Analyzer Air (Carrier)

Purity: Breathing grade; dry. Hydrocarbon
   content less than 10 ppm. (Small amounts
   will be removed by catalytic oxidizer.)
Regulator: Elastomer diaphragm may be used,
   but should be LOX-service cleaned.

Hydrogen (Carrier and Fuel)

Purity: Must be free of methane and CO.
   Dew point -90° F. Purity requirements
   are most stringent for this gas. Methane,
   the most common impurity, will subtract by
   an equivalent amount from the methane
   measurement of the ambient air. Tank
   hydrogen should be specified "electrolytic"
   grade to avoid methane contaminant. However,
   even cylinders marked "electrolytic" have
   been found to contain considerable amounts
   of methane. A palladium-diffusion purifier
   (Trienco) may be used to remove all suspected
   impurities from cylinder hydrogen. Alternatively, an electrolytic hydrogen generator
   (Trienco, Milton Roy) may be used, but its
   output should be checked for water content
   to make sure it is within the specified dew
   point of -90° F. Generators have been known
   to malfunction and to generate wet hydrogen.

Regulator

Cylinder Hydrogen: Metal diaphragm, LOX-service
   cleaned regulator required.
Cylinder Hydrogen with Palladium Diffusion
   Purifier: Elastomeric diaphragm may be used,
   but LOX-service cleaning required to prevent
   possible impurities that could poison
   palladium membrane.

Calibration Gases

Calibration should be done on the highest range practicable. The CO, methane, and the total hydrocarbon readings should be calibrated with a blend containing *methane and CO only* in air, and *must* be free of heavy hydrocarbons. The amount of CO and methane should be such as to give approximately a scale reading of 80% on the *next higher range* above the most sensitive range to be used. That is, of 0-5 ppm range is the most sensitive range to be used, then the blend should contain 35-40 ppm. Calibration is then performed on the next higher range or the 0-50 ppm range. This automatically places the next lower 0-5 ppm range and the next higher ranges 0-500 and 0-5,000 ppm in correct calibration. This is accomplished by the precision (0.1%) range resistors, which provide exactly a factor of 10 between adjacent ranges.

Calibration of ethylene and acetylene on the five-component system requires a separate calibration blend in air background. Concentrations should be an order of magnitude higher as in the case of methane and CO.

The advantages of using higher concentrations for calibration are:

1. Higher concentrations are available with greater accuracy and closer tolerances than low concentrations.
2. The effect of trace heavy hydrocarbons on the accuracy of the total hydrocarbon calibration is minimized. (Heavy hydrocarbons, if present in unknown amounts, will introduce a serious error into the total hydrocarbon measurement by causing too much attenuation to be introduced into that calibration. The result will be too low a reading on ambient samples and can result in apparent negative values of nonmethane hydrocarbons.)

Regulator

Metal diaphragm and LOX-service cleaning required. It is of utmost importance to avoid introducing *any* form of hydrocarbon impurity into the calibration blend. Elastomeric diaphragms are excellent sinks for, and sources of, heavy hydrocarbons.

Zero Gases

A "zero" gas is *not required*. The very nature
of the chromatographic technique is such that a zero
reference is provided by the base line. Moreover,
the technique used for the total hydrocarbon measurement, namely catalytic oxidation of hydrocarbons in
the air carrier, provides a zero-reference for that
measurement. Because this is contrary to the
requirements of almost every other kind of analyzer
in use in air pollution monitoring, numerous attempts
have been made to use "zero" gases to check "zero"
on the total hydrocarbon measurement. Because the
instrument has a sensitivity on the total hydrocarbon
measurement of about 0.2 ppm full scale, we have yet
to find a zero gas free of hydrocarbons that does
not give a significant response on the total hydrocarbon channel. This has proven somewhat discon-
certing to users who, expecting the instrument to
read zero, find instead that it indicates almost
full scale on the most sensitive range. Rasmussen[10]
has reported finding significant quantities of heavy
hydrocarbons in quality zero gases and methane
calibration standards.

Training and Maintenance

Gas chromatographs with flame ionization detectors have been widely used in the hydrocarbon
processing industries for almost 15 years. Although
many such instruments are complex—indeed, many
considerably more so than the air quality chromatograph under discussion—the industry has developed
the capability of maintaining them with a high level
of reliability and a minimum of down time. This has
been possible only because users have recognized
that a highly competent level of maintenance is
required to achieve the desired performance. The
need for proper training and schooling in the techniques and instruments has been recognized and
accepted by users and manufacturers alike. In line
with this, most process analyzer manufacturers offer
maintenance training courses for users of their
equipment.

Similarly, users of air quality chromatographs
will not be disappointed in the performance of these
instruments if their maintenance personnel are
properly trained in the operation and maintenance
of the chromatograph. Ortman[11] has indicated the
following elements essential to a successful installat.

1. field start-up by manufacturers' service
   personnel
2. a minimum of one week's training for users'
   maintenance personnel
3. availability of field service and spare parts
4. advisability of maintenance contract with the
   manufacturer.

Most manufacturers provide these services and
capabilities.

A tuition-free school in maintaining and servicing
the air quality chromatograph is offered by Beckman.
Most users have recognized the value of this school
and have taken advantage of it by enrolling their
personnel who will be responsible for **maintaining**
and servicing the instruments.

APPLICATIONS OF AUTOMATED
AIR MONITORING CHROMATOGRAPHS

The preceding section dealt with the current
status of the air quality chromatograph in the field
of application for which it was designed: the
measurement of nonmethane hydrocarbons and carbon
monoxide in ambient air. As indicated previously,
the automatic capability of the instrument has been
recognized by other workers for its potential use
in other applications. Like any chromatograph, it
can be adapted to a wide variety of measurements and
analytical problems simply by using different columns
and operating conditions. It is not surprising then
that a number of these have been applied to measurements somewhat different from its original use. The
following cases illustrate a number of other applications and potential uses.

Monitoring Individual Hydrocarbons
in Ambient Air

An obvious extension is monitoring individual
hydrocarbon pollutants heavier than methane and C_2's.
Stephens and Burleson[12] sampled ambient air for
analysis by laboratory chromatography. Twenty-six
individual paraffins, olefins, diolefins, and
acetylenes in the C_1 to C_6 range were individually
determined.

They showed that the more reactive components
were greatly reduced in concentration in the afternoon

air as compared to the samples taken early in the morning. Moreover, they established a correlation between the concentrations of methane and that of ethane and propane, and they hypothesized a natural gas contribution to the hydrocarbon content of the ambient air. They also showed that concentrations of these fluctuated rapidly as shown by a series of 11 samples spaced out over a 2½-hour span. Thus, it was possible to identify from the olefinic components the contribution of automotive exhaust, in spite of the presence of much greater concentrations of natural gas components.

Similarly, in the Los Angeles Basin, seepage from oil fields contributes to the total hydrocarbon burden of the atmosphere. This can be distinguished easily from the components contributed by automotive exhaust, as illustrated by the figures that follow.

During final testing of a chromatograph that had been modified to monitor methane, ethylene, ethane, propane, and isobutane, the instrument was allowed to run overnight, sampling the laboratory air. An unusual pattern of concentrations appeared during the night as evidenced by the bargraph record the following morning. The following night the test was repeated, sampling ambient air taken through a probe at some 30 feet above ground level outside the laboratory. The same pattern appeared on the record the next morning. Figure 1.3 shows a section of bargraph at 8 to 9 p.m. As is apparent, the saturated

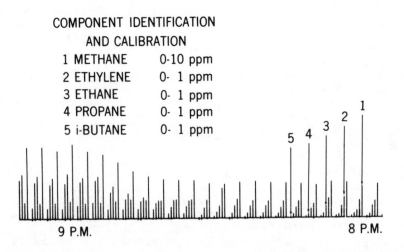

Figure 1.3.   *Bargraph record - ambient air; 30 feet above ground level; 7:40-9:20 p.m., September 20, 1972.*

hydrocarbons began to increase about 8:30 p.m.   By
contrast, the ethylene, which had been at higher
levels during the day, generally drifted downward
in concentration.   A section of bargraph taken the
following morning from 6 to 7 a.m. is shown in
Figure 1.4.   The saturated hydrocarbons had peaked

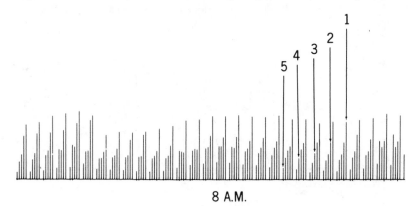

8 A.M.

*Figure 1.4.   Bargraph record - ambient air; 30 feet above ground
level; 7:20-8:50 a.m., September 21, 1972.*

in concentration during the night and now were on the
decrease during the early morning, while ethylene
began increasing due to automotive traffic.   The
concentrations of the five components during the
night have been plotted in Figure 1.5.   Methane,
ethane, propane, and butane are shown in the upper
half and the generally similar trends for each of
these components are apparent.   Ethylene is plotted
separately in the lower section to show the marked
contrast in pattern.   While the saturated hydrocarbon
is low during the day, a sharp increase occurs at
about 8 p.m.   This represents a shift in wind direction from the east to a prevailing wind from the
west, an area where extensive oil fields lie.   By
contrast, ethylene decreases during the night as
traffic subsides and increases sharply during the
morning automobile traffic hours.
    Figure 1.6 shows a manual chromatogram made at
about 8:30 or at the conclusion of the test run
shown in Figure 1.5.   The presence of a considerable
amount of *n*-butane and significant amounts of pen-
tanes is unequivocally indicated.   Moreover, traces
of hexanes seem to be present in amounts just above
the noise level of the chromatograph.   These are

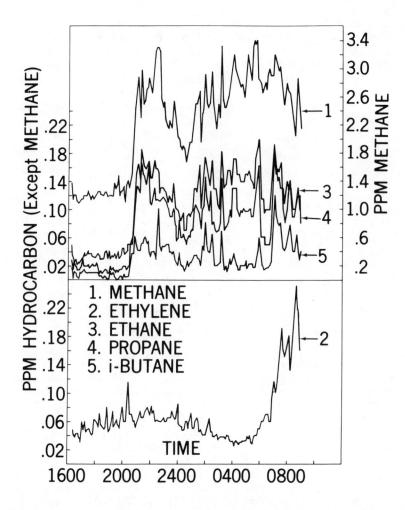

Figure 1.5.   Night-time ambient air hydrocarbons; September
              20-21, 1972; 30 feet above ground level.

estimated to be on the order of 10 to 20 parts per
billion for each of the isomer peaks shown.
    These runs illustrate the value of extended
chromatographic monitoring of ambient air hydrocarbons in helping to assess the contribution of
various sources to the total air pollution. In
areas such as the Los Angeles Basin when contributions of natural sources are significant factors,

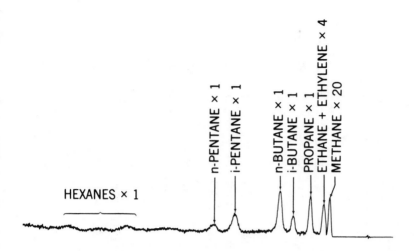

Figure 1.6. Chromatogram - ambient air; 30 feet above ground level; 9:00 a.m., September 21, 1972.

this type of analysis may play an integral part in programs to develop models for the area and the strategies to be used in overcoming air pollution.

Monitoring Industrial Atmospheres

Requirements of the Occupational Safety and Health Act (OSHA) impose stringent rules on the control of toxic chemicals and their release into industrial atmospheres. Employers may find themselves faced with the necessity of continuous monitoring of atmospheres where the presence of toxic compounds is a possibility. In a situation where one or more of a variety of compounds may be present at various times or from various sources, the air quality chromatograph provides a convenient and effective way of monitoring and measuring each individually.

Figure 1.7 shows a chromatogram separation performed on an instrument recently tested and delivered. Columns and operating conditions were selected to provide a repetitive analysis of methane, ethylene, 1,3-butadiene, cyclohexane, and benzene. Minimum full scale ranges obtainable were on the order of better than 1 ppm full scale for each of these compounds, with a minimum detectability on the

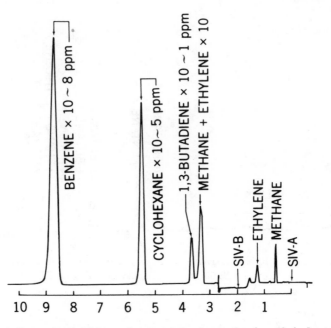

Figure 1.7.   *Air quality chromatograph analysis of industrial atmosphere.*

order of 20-40 parts per billion. The chromatographic system consisted of two separate and independent column systems. One column system separated methane and ethylene. Heavier components were back-flushed to a separate vent. A second column system separated butadiene, cyclohexane, and benzene. Air carrier gas averted the detector upset that occurs with other carriers such as argon or helium. The catalytic oxidizer is used to remove traces of hydrocarbons from the air carrier that would interfere with the measurements.

Similar systems have been designed for monitoring carbon tetrachloride and trichloroethylene. The FID has a relatively poor sensitivity for carbon tetrachloride, but even so, a full scale sensitivity of better than 100 ppm is possible, with a minimum detectability of about 2 ppm.

Smog Chamber Studies

Application of chromatographic analysis in the study of photochemical hydrocarbon reactions in irradiated environmental chambers has been extensive. It has been of particular value in measuring relative reactivity of individual hydrocarbons, as demonstrated by Bellar[13] and co-workers. However, little is known about synergistic effects of mixtures of hydrocarbon because of the difficulties in assessing the effect of other variables on reaction rates, as pointed out by Altshuller.[14] This situation suggests that relatively detailed information about reaction rates in complex hydrocarbon systems can be obtained with an automated air monitoring chromatograph modified to measure the various components individually, and at relatively frequent intervals.

Villalobos and Harman[15] used an air quality chromatograph in an irradiated smog chamber study of the effect of carbon monoxide on reaction rates of ethylene and propylene. Methane, ethylene, propylene, and CO were monitored and recorded during the course of runs up to six hours in length, with a complete analysis every five minutes. Figure 1.8 (from Reference 15) shows a typical run with ethylene, propylene, and NO at a nominal beginning concentration of 10 ppm. Since each hour represents 12 data points for each component, a detailed history of the concentration of each component can be established.

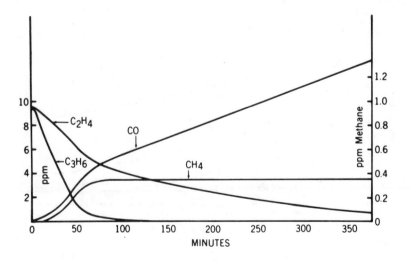

Figure 1.8. Photochemical reaction of ethylene and propylene.

This and other similar runs revealed some interesting aspects of the photochemical reaction of the hydrocarbon reaction and the reaction products.

1. Methane is a product of the photochemical reaction. Its appearance occurs entirely during the early period of the reaction when the rate of CO appearance is the greatest. It is not generated to any measurable extent during the later stages in the reaction.
2. Carbon monoxide's rate of appearance is greatest during the early part of the photochemical reaction.
3. CO continues to be generated after most of the hydrocarbon has disappeared, suggesting that CO is generated as a result of the reaction of a long-lived intermediate product that is present long after most reactants have disappeared.
4. Synergism in the photochemical reactions of ethylene and propylene is strongly suggested.

When the reaction rates of ethylene and propylene when present together were compared to the reaction rates when present singly, a marked difference was observed, as shown in Figure 1.9 (from Reference 15).

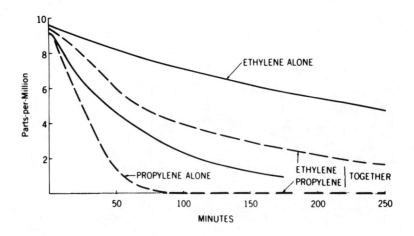

Figure 1.9.   *Synergism of ethylene-propylene photochemical reaction.*

As pointed out by Altshuller, this comparison may not be valid because of the differences in NO_x/hydrocarbon ratio under the two conditions. However, if the reaction rates as determined from the slopes of the lines in Figure 1.9 are plotted, the plot shown in Figure 1.10 is obtained. These data confirm the synergism suggested by Figure 1.9.

Because of the small size of the smog chamber used in this study (72 liter) and the large surface-to-volume ratio obtained, the results may not be a true measure of the "real" world. Nevertheless, the results validly demonstrate the feasibility of accurate measurements of this type by obtaining a density of data possible only with an automated chromatographic system.

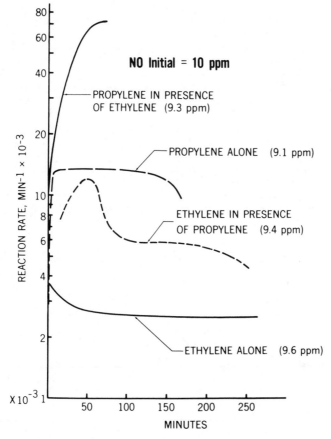

Figure 1.10. *Synergism effect on reaction rates of ethylene-propylene photochemical reaction.*

Automobile Exhaust Emission Measurements

Because the automobile is the major source of
hydrocarbons in urban atmospheres, considerable
attention has been placed on the measurement and
control of unburned hydrocarbons in automotive exhaust. Current EPA standards for 1975 and later
years are so stringent as to bring into question
the ability of the automobile industry to achieve
compliance. Because of the nonreactive nature of
methane, a substantial constituent in automotive
exhaust, some thought has been given to the measurement of nonmethane hydrocarbons in exhaust as a
measure of its polluting potential.

A batch analytical technique, such as the gas
chromatograph, is not ideally suited for end-of-the-
assembly-line testing in automobile manufacturing.
However, it is suitable for testing in accordance
with federal standards using the constant volume
sampling (CVS) method. In this test, an automobile
is run through a series of modes (cold start) and
the exhaust is diluted with ambient air to provide
a constant total volume. A portion thereof is collected in a plastic bag for subsequent analysis.
The test is repeated under slightly different conditions (hot start) and a second bag sample is collected
and analyzed. A third bag collects a sample of the
diluting air that is also analyzed to give a blank
correction. The total time required for the test
is not less than 22 minutes.

Required measurements for each bag and the
conventional instruments used are as follows:

Measurement	Conventional Instrumentation
CO	NDIR
Hydrocarbons	FIA or NDIR
CO_2	NDIR
NO_x	CLA

NDIR = Nondispersive Infrared Analyzer
FIA = Flame Ionization Analyzer
CLA = Chemiluminescent Analyzer

CO_2 is needed in order to calculate the dilution by comparison with a measurement of the raw exhaust as determined by a continuous analyzer. No method in use can determine the nonmethane hydrocarbons directly.

As will be noted, four separate analyzers are required to make these measurements. However, a single chromatograph can be modified to make all these measurements with the exception of NO_x. Figure 1.11 shows a chromatogram produced with an

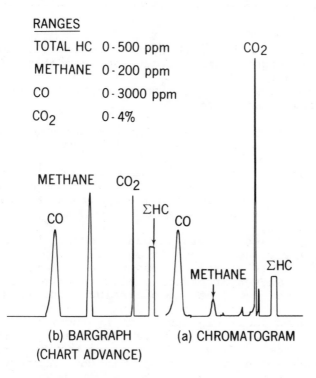

RANGES

TOTAL HC   0-500 ppm
METHANE   0-200 ppm
CO   0-3000 ppm
CO_2 0-4%

Figure 1.11.   *Auto exhaust emission measurement by gas chromatograph CVS testing.*

air quality chromatograph modified to make these measurements. The following ranges were used:

Measurement	*Full Scale Concentration*
Total hydrocarbons	500 ppm
CH_4	200 ppm
CO_2	4%
CO	3000 ppm
Nonmethane hydrocarbons	by difference

The five-minute analysis cycle permits analysis of the three bags within a time frame compatible with the 22-minute test.

The chromatograph was designed to meet the following objectives:

1. short time cycle compatible with the CVS cycle
2. capability for measuring total hydrocarbons, methane, CO, and CO_2
3. total hydrocarbon measurement data to be compatible and correlatible with continuous flame ionization analyzers previously used for this measurement.

The third objective was deemed necessary because of the extensive use of continuous flame ionization analyzers (FIA's) for the measurement of total hydrocarbons. Such hydrocarbon analyzers have been used widely under a variety of operating conditions and with a variety of fuels. Hydrogen/helium (40/60) fuel is most commonly used, but hydrogen/nitrogen (40/60) and pure hydrogen are also used. A wide range of sample flow and combustion air flow rates are also used by various users in the industry. Hence, it was important that the chromatograph be able to operate over the whole range of conditions used for the FIA's so that data previously obtained with the FIA's would be comparable with the chromatograph.

To perform this analysis the chromatograph was modified as follows. A third channel was added for the measurement of CO_2 at the high concentrations encountered. It consisted of a column to separate the CO_2 from the methane and ethane; back-flushing water and all components heavier than CO_2. A small sample size (20 microliters) brought the CO_2 within the linear range of the detector. Furthermore, the system was designed to permit operation on two

chromatograph channels with any of the three above-
named fuels with a combined flow rate in both
channels of about 40-50 ml/min. Make-up fuel is
controlled by a regulator so that additional fuel
to bring the fuel rate up to any desired flow rate
from 50-100 ml/min may be added. Similarly, the
total hydrocarbon channel can be operated for a
carrier flow rate of from 1-25 ml/min, which is the
range of sample flow rates used with the FIA's.

The flame ionization detector burner used in the
air quality chromatograph is identical to that in
the continuous flame ionization analyzers; hence, it
would be expected that operating both instruments
under identical conditions should result in identical
response characteristics. This was tested at
several conditions typical of modes of operation
used for FIA's by the automobile industry, and
listed in Table 1.1. The response of a number of
hydrocarbons was determined for both instruments
with dilute samples of the pure hydrocarbon in air
background at approximately the 100 ppm level. The
results are shown in Table 1.2.

Table 1.1

Test Conditions Response to Various Hydrocarbons
Air Quality Chromatograph vs. Flame Ionization Analyzer

Fuel	I H_2/He	II H_2/He	III H_2/N_2
Fuel rate, ml/m			
AQ/GC	116	67	67
FIA	110	66	65
Support air, ml/m			
AQ/GC	400	260	260
FIA	400	260	260
Sample flow, ml/m			
AQ/GC	23	5.2	5.2
FIA	23	5.5	5.5
Response ratio AQGC/FIA			
Methane	2.45	3.43	2.95

Table 1.2

Relative Response to Various Hydrocarbons
Air Quality Chromatograph vs. Flame Ionization Analyzer

Fuel	I H_2/He	II H_2/He	III H_2/N_2
Relative response ratio AQGC/FIA			
Methane	1.00	1.00	1.00
Ethane	1.01	0.99	1.00
Propane	1.01	0.98	1.00
n-Butane	1.01	0.98	1.01
n-Hexane	1.00	0.97	0.98
Ethylene	1.01	0.97	0.99
Propylene	0.99	0.98	0.98
Butene-1	1.00	1.01	0.98
Butadiene-1,3	1.01	0.96	0.99
Acetylene	1.03	1.12	1.07
Methylcyclopentane	1.00	0.97	0.98
Benzene	1.01	0.98	1.00

The relative response ratio was determined by analyzing the samples simultaneously with both instruments. Absolute response ratio for each compound was obtained by ratioing the chromatograph reading to the FIA reading. The relative response ratio was then obtained by ratioing the absolute response ratio for each compound to the absolute response ratio for methane. The results indicate that when both instruments are calibrated against the same compound (in this case, methane), they agree remarkably well. With the exception of acetylene, which deviates as much as 12%, most components agree to within 2 or 3%. This indicates that the chromatograph, when operated under the same conditions as the FIA, will give the same results for the total hydrocarbon analyzer.

Although the chromatograph is higher in initial cost than the continuous FIA, it replaces not one but three instruments: the FIA for total hydrocarbons and two NDIR's for measuring CO and CO_2 respectively. Overall investment is thereby considerably reduced, with the additional advantages of the nonmethane hydrocarbons measurement.

Stack Gas Monitoring

Although federal standards do not require the measurement of carbon monoxide, carbon dioxide, or unburned hydrocarbon emissions from stacks, nevertheless, studies directed at optimizing combustion processes will undoubtedly benefit from the availability of automated chromatographs. Such a unit has been placed in service in a mobile instrument trailer to monitor CO, CO_2, and nonmethane hydrocarbon content of stack gases. It has been reported that the chromatograph is used in the conventional manner: the stack gas sample is drawn through a heated line, through a cooler to remove excess water, and then into the chromatograph. It is hoped that more complete data regarding this operation will be available at a later date.

Dissolved Hydrocarbons in Seawater

Dissolved hydrocarbons in the ocean do not represent a serious air pollution problem, at least not at normal background levels of concentration. Nevertheless, this application of the air quality chromatograph is of sufficient novelty to merit describing as much as is known about it.

It is known that the location of undersea oil pools is indicated by the presence of dissolved hydrocarbons in the seawater in the vicinity of the pool, at concentrations greater than the normal background existing in ocean water. To detect this above-normal concentration, seawater is brought up from the desired level by means of a sampling probe towed by the ship. It is then passed through an extraction apparatus that quantitatively transfers the hydrocarbons to a gas stream that is subsequently analyzed by a gas chromatograph.

Figure 1.12 shows a chromatogram as produced by an instrument designed to these specifications. With two separate column systems, both using air carrier, the instrument was able to make an analysis of methane, ethane, ethylene, propane, iso-butane,

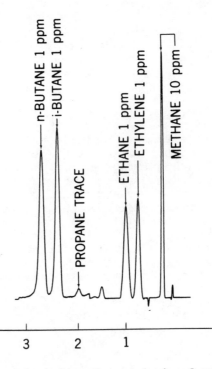

Figure 1.12. *Light hydrocarbon analysis of seawater extract, calibration standard*

and normal butane on a three-minute cycle. One column system separated methane, ethane, and ethylene during the first minute of the cycle, while the second separated the rest in an additional two minutes. Sensitivity obtained is somewhat better than 1 ppm full scale for all components. This provided a minimum detectability of approximately 20 parts per billion for each component.

SUMMARY AND CONCLUSIONS

Ambient air quality monitoring gas chromatographs have been developed for continuous, unattended monitoring of methane, total hydrocarbons, and carbon monoxide in ambient air at levels as low as 1 ppm. These instruments have been designed to meet federal

standards for the measurement of nonmethane hydrocarbons. Use of these instruments is increasing,
and it is expected that they will be extensively used
in ambient air monitoring stations to be built and
equipped during the next few years as state and local
agencies implement their strategies to reduce pollution and meet federal clean air standards. Since
their introduction, a number of improvements have
been made to obtain better performance, increased
reliability, and reduced maintenance. These
improvements have been described.

In addition, field experience to date indicates
the need for increased attention to (1) the quality
of the installation, (2) the purity of the supply
gases, and (3) the technical training level of maintenance personnel. Suggested criteria based on past
experience have been outlined.

Automatic ambient air quality chromatographs
have also found application in a variety of uses
quite different from those for which they were
generally designed. A number of these applications
have been reviewed with the objective of demonstrating
the adaptability and flexibility inherent in the gas
chromatograph and to demonstrate the advantages
derived from their automated nature and their
capability of providing large amounts of detailed
data.

REFERENCES

1. Stevens, R. K., A. E. O'Keefe, and G. C. Ortman. "A Gas
   Chromatographic Approach to the Semi-Continuous Monitoring
   of Carbon Monoxide and Methane." Presented at the 156th
   National ACS Meeting, Atlantic City, New Jersey (1968).
2. Stevens, R. K., A. E. O'Keefe, and G. C. Ortman. "Current
   Trends in Continuous Air Pollution Monitoring Systems,"
   ISA Trans., 9(1), 1 (1970). Also see *Anal. Chem., 42(2),*
   142a (1970).
3. Schwink, A., H. Hockenberg, and Mr. Forderreuther.
   "Spurensuche mit Flammen ionisations-detektor und depakten
   Trennsaulen in der Gas chromatographic II," *Brennstoff-
   Chemic, 42(9),* 295 (1961).
4. Porter, K. and D. H. Volman. "Flame Ionization Detection
   of Carbon Monoxide for Gas Chromatographic Analysis,"
   Anal. Chem., 34, 748 (1962).
5. Johns, T. and B. Thompson. "Analysis of Inorganics and
   Selective Detection with a Flame Ionization Detector."
   presented at the Pittsburgh Conference on Analytical
   Chemistry and Applied Spectroscopy, March 1965.

6. DuBois, L., A. Zdrojewski, and J. L. Monkman. "The Analysis of Carbon Monoxide in Urban Air at the ppm Level, and the Normal Carbon Monoxide Level," *J. Air Poll. Control Assn., 16*, 135 (1966).

7. Altshuller, A. P., S. L. Kopczynski, W. A. Lonneman, T. L. Becker, and R. Slater. "Chemical Aspects of the Photooxidation of the Propylene-Nitrogen Oxide System," *Environ. Sci. Technol., 1*, 899 (1967).

8. Villalobos, R., and R. L. Chapman. "A Gas Chromatographic Method for Automatic Monitoring of Pollutants in Ambient Air," *ISA Trans., 10(4)*, 356 (1971).

9. Villalobos, R., D. Stevens, R. LeBlanc, and L. Braun. "Design and Performance of a Gas Chromatograph for Automatic Monitoring of Pollutants in Ambient Air," AIAA Paper No. 71-1065. Presented at the Joint Conference on Sensing of Environmental Pollutants, Palo Alto, California (November 8-10, 1971).

10. Rasmussen, R. Presented at ASTM Committee E-19, 11th Annual Meeting on "The Practice of Chromatography," September 24, 1972, St. Louis, Missouri.

11. Ortman, G. C. Presented at ASTM Committee D-22, "Symposium on Present Status of Air Quality Instrumentation," October 25, 1972, Philadelphia, Pennsylvania.

12. Stephens, E. R. and F. R. Burleson. "Distribution of Light Hydrocarbons in Ambient Air," *J. Air Poll. Control Assn., 19*, 929 (1969).

13. Bellar, T., J. E. Sigsby, C. A. Clemens, and A. P. Altshuller. "Direct Application of Gas Chromatography to Atmospheric Pollutants," *Anal. Chem., 34*, 763 (1962).

14. Altshuller, A. P. and J. J. Bufalini. "Photochemical Aspects of Air Pollution: A Review," *Environ. Sci. Technol., 5(1)*, 39 (1971).

15. Villalobos, R. and J. N. Harman III. "Use of an Automated Gas Chromatograph in Smog Chamber Studies." Presented at ISA Analysis Instrumentation Division Symposium, April 23-26, 1973, St. Louis, Missouri.

16. Westberg, K., N. Kohn, and K. W. Wilson. "Carbon Monoxide: Its Role in Photochemical Smog Formation," *Science, 171*, 1013 (1971).

CHAPTER 2

APPLICATION OF CHEMILUMINESCENCE TO THE MEASUREMENT OF GASEOUS POLLUTANTS

J. A. Hodgeson, W. A. McClenny and R. K. Stevens

Chemiluminescence and bioluminescence systems* have long been of considerable scientific interest. The emission obtained is characteristic of reaction products or excited intermediates and thus provides a natural means for studying reaction mechanisms. Measurement of chemiluminescent intensities has been used to provide concentration measurements in kinetics studies. However, conventional analytical applications of chemiluminescence have only appeared over the past decade. Beginning in 1968, research and development in the area of chemiluminescence detectors for air pollutants was actively initiated. At present such detectors are being applied routinely for monitoring atmospheric concentrations of ozone, oxides of nitrogen and sulfur dioxide.[1,2]
Two types of chemiluminescence detectors have been applied in air pollution measurements. The ambient temperature detector (Figure 2.1) employs the chemiluminescence reaction between the molecule of interest in air and a second reactive species. Another type of detector is one in which molecular chemiluminescence emissions are observed in a cool flame (Figure 2.2). Whereas the first type of detector is used for the specific measurement of a single pollutant, the flame detector is more applicable to the detection of classes of compounds, *e.g.*, gaseous sulfur compounds.

---

*Mention of a company or product name is not intended to constitute endorsement by the Environmental Protection Agency.

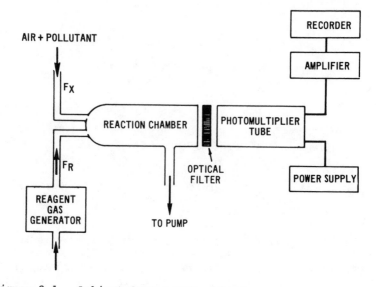

Figure 2.1.   Ambient temperature detector.

The application of chemiluminescence has provided a class of emission detectors that have some common characteristics. Among these are an inherent degree of sensitivity, specificity and simplicity. Quantitative measurements down to one part per billion (1 ppb) of pollutant in air are quite feasible. The chemiluminescence reaction alone is quite selective and the use of optical filters further enhances specificity. The detector usually consists of gas inlet, reaction zone, air pump, optical filter, photomultiplier tube and associated electronics. The applicability of the general method is limited to that class of molecules that enter directly, or indirectly, into chemiluminescence reactions. Another general limitation is the nonabsolute nature of the measurement. Chemiluminescence detectors are normally calibrated with known concentrations of gas samples.

The discussion below will provide a summary of recent research, development and application in the area of chemiluminescence analysis for ozone, oxides of nitrogen and sulfur compounds. Commercially available chemiluminescent air pollution monitors are tabulated, and recent developments that promise to extend the applicability of chemiluminescence analysis are discussed.

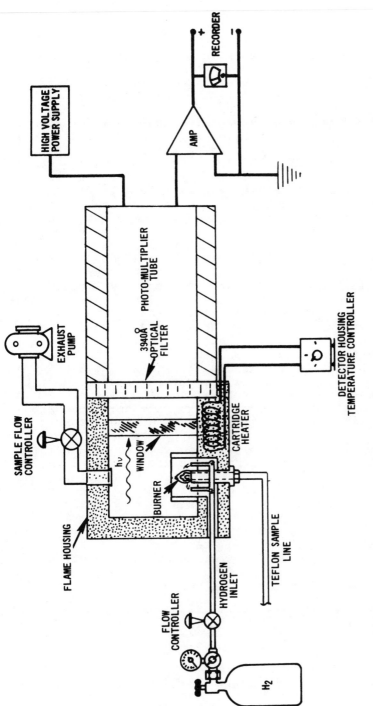

Figure 2.2. Typical commercial flame photometric detector.

OZONE

Chemiluminescence is a characteristic feature of the reactions of ozone with many inorganic and organic materials.[3] Regener[4] utilized the intense emission from the reaction between ozone and Rhodamine B, adsorbed on an activated silica gel surface, in the first chemiluminescence detector for atmospheric ozone. The response characteristics of the detector were studied and a chemiluminescence surface with improved lifetime and stability was developed.[5] An automated version of the Regener analyzer is commercially available from Tritek Corporation, Chapel Hill, North Carolina. Ozone measurements by the Regener procedure have the advantage of requiring no support gases and only periodic replacement of the Rhodamine B surface. Because the surface sensitivity changes, frequent recalibration is required, and the electronics and gas-flow arrangements are somewhat complicated.

A homogeneous gas phase chemiluminescence technique for the detection of O_3 was reported by Nederbragt.[6] The detector employed an atmospheric pressure chemiluminescent reaction between ozone and ethylene. The reaction between ozone and ethylene yields chemiluminescence emission in the 300-600 nm region (λ max $\simeq$ 435 nm).[7] Prototype detectors based on Nederbragt's concept were constructed and evaluated by EPA and were shown to have more than adequate sensitivity and specificity for ambient ozone measurements.[7] The Nederbragt method has subsequently been so successful in laboratory and field applications that it has been designated as the reference method for the routine ozone measurements required by recent federal air quality standards.[8] Several instrument companies are now offering commercial versions of this detector (Table 2.1). In a period of only 18 months, the Nederbragt approach has evolved through the stages of laboratory prototype construction and evaluation, field evaluations and comparisons to promulgation as a reference method and commercialization.

SULFUR COMPOUNDS

At this date, flame chemiluminescence is the only approach being used routinely for monitoring atmospheric sulfur compounds. When volatile sulfur compour are burned in a hydrogen rich flame, an intense blue

Table 2.1

Chemiluminescent Ozone Monitors*

	Monitor Model No.
Bendix Instrument Division Ronceverte, West Virginia	8000
Beckman Instruments, Inc. Fullerton, California	950
Rem Incorporated Santa Monica, California	612B
McMillan Electronics Corporation Houston, Texas	1100-2B
Meloy Labs, Incorporated Springfield, Virginia	300
Kimoto Electric Company Osaka, Japan	804

*There may be manufacturers of chemiluminescent air pollution monitors other than those listed in Tables 2.1-2.2 and mentioned in the text; any omission of manufacturers is unintentional.

chemiluminescence occurs. This emission results from the recombination of sulfur atoms formed in the reducing flame environment. The chemiluminescence consists of a series of evenly spaced bands between 350 and 450 nm. A photomultiplier views the intense band at 394 nm through an interference filter. The chemiluminescence intensity is proportional to the square of the sulfur concentration (for molecules containing one sulfur atom) since the reaction involves the recombination of two sulfur atoms. The minimum detectable limit of current flame photometric detectors is 0.005 ppm of sulfur compound.

The original flame photometric detector (FPD) for sulfur compounds was revealed in a patent by Draeger.[9] Brody and Chaney[10] described the application of this detector to gas chromatography. The calibration and application of the FPD for detecting atmospheric concentrations of SO_2 was discussed by Stevens, *et al.*[11] Meloy and Bendix Corporation both offer the flame photometric detector as a sulfur dioxide monitor. For areas where hydrogen sulfide may coexist with SO_2, they offer a scrubber as an accessory, usually silver-heated to 135°C to remove hydrogen sulfide without removing SO_2.

Stevens, *et al.*[12] have extended the specificity of the FPD by coupling a special gas chromatographic (GC) column to the detector. A GC column was developed that is capable of separating and quantitatively eluting atmospheric concentrations of individual sulfur compounds, including SO_2, H_2S, simple mercaptans and organic sulfides. The flame photometer is used as the detector on the column effluent. Commercial sources of the FPD-GC sulfur compound monitor are listed in Table 2.2.

Table 2.2

FPD-GC Sulfur Chromatographs

	Model No.
Analytical Instrument Development Company West Chester, Pennsylvania	513
Bendix, Process Instrument Division Ronceverte, West Virginia	8700
Tracor Austin, Texas	270
Varian Aerograph Walnut Creek, California	1490-5

In studies on potential atmospheric degradation of nonpolluted areas by encroachment of industrialization or urbanization, knowledge of pollutant background concentrations is required. In the case of SO_2, the ambient background may be much less than 0.01 ppm. Commercial flame photometric detectors have inadequate sensitivity for such measurements. Recent studies have applied correlation techniques to improve sensitivity for SO_2 detection.[13,14] The flame emission spectrum of SO_2 is a banded structure, and the conventional FPD views only the wavelength of a band maximum. The photomultiplier thus sees the sum of sulfur emission and flame background.

In the correlation approach, one or more band maxima are viewed and compared to adjacent background wavelengths. By peak and valley comparison, cancellation of the flame background term is feasible.

Figure 2.3 illustrates a simple flame correlation
detector for SO_2. Two interference filters, which
separately view a band maximum and an adjacent
background wavelength, are mounted on a rotating
blade. The photomultiplier alternately sees the
peak emission and the emission from the flame background alone. The lock-in amplifier is used to
demodulate the alternating signal and provide an
output corrected for flame background.

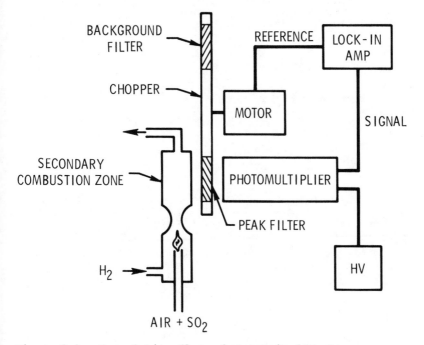

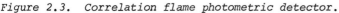

Figure 2.3. Correlation flame photometric detector.

Ambient temperature chemiluminescence reactions
have been suggested for measurement of SO_2 and other
sulfur compounds, but applications have not yet been
reported. Snyder and Wooten[15] investigated the use
of O-atom chemiluminescence for the detection of SO_2.
Reaction of SO_2 and O-atoms produces chemiluminescence
in the ultraviolet region.

$$O + O + SO_2 \rightarrow O_2 + SO_2*$$

$$SO_2* \rightarrow SO_2 + h\nu \ (\lambda_{max} = 280 \text{ nm})$$

The sensitivity for SO_2 detection using this chemiluminescence is 0.001 ppm, which is quite adequate
for atmospheric detection. However, practical O-atom
sources having good stability and lifetime have not
yet been developed. The ambient temperature SO_2
chemiluminescence detectors thus have not evolved
beyond the laboratory stage. Finally, Kummer, *et al.*[1]
have observed chemiluminescence from the gas phase
reactions of O_3 with a variety of compounds, including H_2S and organic sulfides, and suggested use
of this chemiluminescence for measuring atmospheric
sulfur compounds.

OXIDES OF NITROGEN

The principal chemiluminescence reaction that has
been used in the detection of oxides of nitrogen is
that between nitric oxide (NO) and ozone (O_3).

$$NO + O_3 = NO_2^* + O_2$$

$$NO_2^* = NO_2 + h\nu \ (0.6 - 3 \ \mu)$$

The direct chemiluminescence reaction is applicable
only to the detection of NO. For detection of
NO_x (NO + NO_2) a prior conversion of NO_2 to NO is
required. Another reaction applicable to the
detection of NO involves atomic oxygen.

$$NO + O + M = NO_2^* + M$$

$$NO_2^* = NO_2 + h\nu \ (0.4 - 1.4 \ \mu)$$

Any NO_2 present is rapidly converted to NO via reaction with O-atoms. The emission intensity is thus
proportional to the sum of NO plus NO_2 (NO_x). This
reaction may be applicable to the direct detection
of NO_x at the high concentrations existing in source
emissions.

The feasibility of detecting ambient concentrations of NO by chemiluminescence was first demonstrat
by Fontijn, *et al.*[17] The earlier detectors used a
small vacuum pump that pulled the air sample and O_3
stream through the reaction chamber at total pressure
of 1 to 5 torr. The detector response time is
approximately 1 sec, and no interferences have been
observed in the chemiluminescent detection of NO in
ambient air. In a recent development the feasibility

of detecting ambient concentrations of NO in a re-
actor cell operated at atmospheric pressure was
demonstrated.[18] Several commercial chemiluminescence
NO_x models are being sold today (Table 2.3), all of
which operate at pressures of 200 torr or greater
and employ small diaphragm-type air pumps.

Table 2.3

Oxides of Nitrogen Monitors

	Model No.
Aerochem Research Labs, Incorporated Princeton, New Jersey	AA-5
Beckman Instruments, Incorporated Fullerton, California	952
Bendix Ronceverte, West Virginia	8100
McMillan Electronics Corporation Houston, Texas	1200
Meloy Labs Springfield, Virginia	NA-520
Rem Incorporated Santa Monica, California	642
Thermo Electron Corporation Waltham, Massachusetts	14B
Scott Research Labs Plumsteadville, Pennsylvania	225
Monitor Labs Incorporated San Diego, California	8440
Yanaco Kyoto, Japan	ECL-7

Present ambient air quality standards only require NO_2 measurements, whereas NO_x measurements are
needed for sources. Because of the enviable features
of the NO chemiluminescence monitor, during recent
years considerable effort has been expended on
development of methods for the conversion of NO_2 to
NO. The earliest work on NO_2 conversion for application to chemiluminescence analysis was by Sigsby,
et al.,[19] who observed quantitative conversion of
NO_2 to NO through a stainless steel tube heated to
temperatures of 600°C. Under these conditions,
however, ammonia (NH_3) is oxidized to NO and becomes
a potential interference. The reduction of NO_2 at
lower temperatures, where no oxidation of NH_3 occurs,
is the method commonly used today for eliminating
the NH_3 interference. Over the past 18 months,
considerable work has been performed within EPA and
industry on the development, improvement and evaluation of various carbon-based converters. The use
of carbon has been demonstrated to be a workable
approach for the specific redox conversion of NO_2.

$$C + NO_2 = CO + NO \; \Delta F° \; 500°K = -32.2 \; Kcal$$

Carbons of various forms, *e.g.*, graphite and carbon
black, are currently being used in commercial monitor
   In a recent development, a direct chemiluminescen
method for NO_2, which employed a "photofragment
detector," was demonstrated.[20] NO_2 is first photolyz
in the 300-400 nm region and the resultant O-atom
detected by chemiluminescent reaction with NO. The
detector does not respond to NO, the response is
linear over the ambient concentration range, and the
lower limit of detection is approximately 0.001 ppm
NO_2.
   A flame photometric detector for nitrogen compounds has recently been reported,[21] which is
analogous in many respects to the FPD sulfur detector
When nitrogen compounds are burned in a hydrogen-rich
flame, emission characteristics of the HNO molecule
were observed in the near-infrared (650-750 nm). By
monitoring this emission in a conventional FPD design
a detector specific for nitrogen compounds, including
NO and NO_2, was obtained. The minimum detectable
limit of the present design is about 0.05 ppm. For
monitoring atmospheric NO_x, the highly specific and
more sensitive ambient temperature detector is pre-
ferred. Provided increased sensitivity can be
achieved, the flame photometer for nitrogen could

be quite useful as a chromatographic detector for organic nitrogen compounds in the atmosphere or other media.

REFERENCES

1. Stevens, R. K. and J. A. Hodgeson. *Anal. Chem.*, *45*, 443A (1973).
2. Fontijn, A., D. Golomb and J. A. Hodgeson. in *Chemiluminescence and Bioluminescence*, M. J. Cormier, D. M. Hercules and J. Lee, Eds. (New York: Plenum Press, 1973), pp. 393-426.
3. Bernanose, H. J. and M. J. Rene. *Advances in Chemistry Series 21* (Washington, D.C.: American Chemical Society, 1959), pp. 7-12.
4. Regener, V. H. *J. Geophys. Res.*, *65*, 3975 (1960); *69*, 3795 (1964).
5. Hodgeson, J. A., K. J. Krost, A. E. O'Keefe and R. K. Stevens. *Anal. Chem.*, *42*, 1795 (1970).
6. Nederbragt, G. W., A. Van der Horst, and J. Van Duijn. *Nature*, *206(4979)*, 87 (1965).
7. Hodgeson, J. A., B. E. Martin and R. E. Baumgardner, Environmental Protection Agency, Research Triangle Park, N.C. Eastern Analytical Symposium, New York, N.Y., Paper No. 77 (1970).
8. Environmental Protection Agency, *Federal Register*, *36(228)*, 22384-22397 (1971).
9. Draeger, B., H. Draegerwerk. W. German Patent 1,133,918, July 26, 1962.
10. Brody, S. S. and J. E. Chaney. *J. Gas Chromat.*, *4*, 42 (1966).
11. Stevens, R. K., A. E. O'Keeffe and G. C. Ortman. *Environ. Sci. Technol.*, *3*, 652 (1969).
12. Stevens, R. K., J. D. Mulik, A. E. O'Keeffe and K. J. Krost. *Anal. Chem.*, *43*, 827 (1971).
13. Horning, A. Contract Report, Environmental Protection Agency, Research Triangle Park, N.C., Contract No. EHSD 71-50 (1972).
14. Shiller, J. W. *Bendix Tech. J.*, *4*, 56 (1971).
15. Snyder, A. D. and G. W. Wooten. Final Report, EPA Contract No. CPA-22-69-8, NTIS PB-188-103 (1969).
16. Kummer, W. A., J. N. Pitts, Jr., and R. P. Steer. *Environ. Sci. Technol.*, *5*, 1045 (1971).
17. Fontijn, A., A. J. Sabadell and R. J. Ronco. *Anal. Chem.*, *42*, 575 (1970).
18. Hodgeson, J. A., K. A. Rehme, B. E. Martin, and R. K. Stevens. 1972 Air Pollution Control Association Meeting, Miami, Florida, June, 1972, Paper No. 72-12.

19. Sigsby, J. E., F. M. Black, T. A. Bellar and D. L.
     Klosterman. *Environ. Sci. Technol.*, *7*, 51 (1973).

20. McClenny, W. A., J. A. Hodgeson and J. P. Bell. *Anal.*
     Chem., *45*, 1514 (1973).

21. Krost, K. J., J. A. Hodgeson and R. K. Stevens. *Anal.*
     Chem., *45*, 1800 (1973).

CHAPTER 3

DIODE LASERS FOR POLLUTANT MONITORING*

E. D. Hinkley and A. R. Calawa

INTRODUCTION

Several optical techniques are presently in use
or under development for *in situ* monitoring of
atmospheric pollutant gases. Their operation is
based upon the detection of characteristic emission
or absorption lines in the visible, ultraviolet, or
infrared regions of the electromagnetic spectrum.
With the advent of tunable lasers, substantially
higher power over narrow, selectable spectral intervals has become available, and with it the potential
for increasing both the sensitivity and specificity
of future monitoring systems.[1]
Because of their small size, ruggedness, and
simplicity of operation, semiconductor diode lasers
have been considered for a wide variety of gaseous
detection applications, from point-sampling to
ambient air measurements.[2] Diode lasers from various
compositions of the semiconductor materials $PbS_{1-x}Se_x$
and $Pb_{1-x}Sn_xTe$ emitting in the 4-12 μm region of the
infrared have been used to measure characteristic
absorption of such gases as CO, NO, NO_2, CH_2O, SO_2,
O_3, NH_3, C_2H_4, and water vapor.
In this paper we will describe the application
of diode lasers to ambient air and source monitoring
by infrared absorption, and give examples from
experiments already performed. We will conclude

---

*This work was supported by the Environmental Protection Agency
and the National Science Foundation (RANN).

with a discussion of some of the engineering obstacl that must be overcome before the use of these device can become widespread.

SEMICONDUCTOR DIODE LASERS

The development of ternary semiconducting compounds of adjustable chemical composition and energy gap[3] permitted diode lasers to be fabricated at nominal wavelengths over a wide range in the infrare Particularly useful have been the Pb-salt compounds; complete coverage of the infrared "fingerprint" region of interest for gaseous pollutant monitoring is possible with only three types: $Pb_{1-x}Cd_xS$, $PbS_{1-x}Se_x$, and $Pb_{1-x}Sn_xTe$. The wavelength span for each of these materials at 10° K is shown in Figure 3.1. As an example, $Pb_{0.88}Sn_{0.12}Te$ lasers emit radiation around 945 cm^{-1}, while $Pb_{0.93}Sn_{0.07}Te$ lasers emit around 1130 cm^{-1}. Lasers made from $Pb_{1-x}Cd_xS$

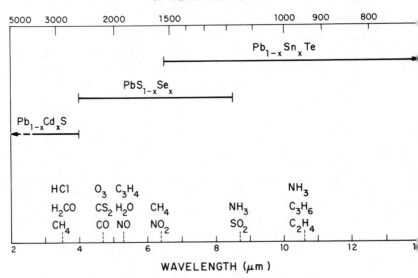

Figure 3.1. *Nominal wavelength regions covered by three Pb-sa semiconductors at 12° K. Strongly-absorbing wave lengths of some common gaseous pollutants are als indicated.*

crystals should be particularly useful for monitoring the hydrocarbons, which all have unique C-H bond absorption spectra in the 3-4 μm region.

Figure 3.2 shows a diode laser in its standard package, which is approximately the size of an average transistor. The laser crystal itself is

Figure 3.2. Photographs of diode laser mounted in standard package. Laser emission is in a direction normal to the page.

nominally 1.2 x 0.4 x 0.2 mm in size, and is shown in the magnified view of Figure 3.2(b). The diode package is mounted onto a copper stud that is usually attached to the "cold finger" of a liquid helium Dewar. (More recently, closed-cycle cryogenic coolers have been used to eliminate the need for liquid helium.) Laser emission is produced when an electrical injection current (either steady or pulsed) is passed through the diode; the wavelength can be simply changed from its nominal value (determined by composition and ambient temperature of the device) by changing the junction temperature—usually by varying the average current level.

MONITORING

Point Sampling

Figure 3.3 consists of a pair of high-resolution laser scans in the 10-μm region of the gases SF_6 and C_2H_4. (The wavenumber scale is referenced to the P(14) CO_2 laser transition by a heterodyne-absorption technique.[4]) At the low pressures used, the spectral

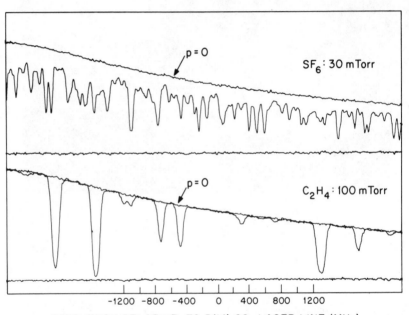

FREQUENCY RELATIVE TO P(14) CO_2 LASER LINE (MHz)

Figure 3.3. Doppler-limited spectra of SF_6 (top) and C_2H_4 (bottom trace) as taken with a $Pb_{0.88}Sn_{0.12}Te$ diode laser. Gas temperature: 293° K, cell length: 30 cm.

lines are Doppler-broadened to widths (FWHM) related to the masses of the individual molecules (29 MHz for SF_6, and 66 MHz for C_2H_4). It is clear that the spectral "signatures" of these two gases are easily distinguishable, and that very high specificity can be achieved under point-sampling conditions where gas pressure is low enough to make collision-broadening negligible. Examples of point sampling using tunable diode lasers have already been reported in the literature.[1,2]

Ambient Air and *In Situ* Source Monitoring

One of the main advantages of using optical techniques for atmospheric monitoring is that sampling is not necessary, and measurements can be performed *in situ*. The spectral lines of gases in the atmosphere are pressure-broadened, however, and this tends to reduce some of the advantages of high sensitivity and specificity obtained at reduced pressure (see Figure 3.3).

Line Spectra at Atmospheric Pressure

An example of the infrared absorption structure of SO_2 in the atmosphere is shown in Figure 3.4 for a portion of the ν_1 band. This spectrum was generated by a computer using the fundamental lines whose positions were obtained on the basis of high-resolution diode laser scans,[5] and whose widths (0.3 cm^{-1}) were also measured directly by diode laser spectroscopy. This spectrum compares favorably with recent experimental data of Burch *et al*,[6] taken with a high-resolution (~ 0.1 cm^{-1}) grating spectrometer. Each of the strong absorption peaks, such as those labeled *A* and *B*, is comprised of many discrete lines that overlap at atmospheric pressure. For CO, NO, and other gases having spectral lines separated by 1 cm^{-1} or more, individual lines can easily be resolved at atmospheric pressure. The total compositional tuning range of diode lasers is approximately 4300 cm^{-1} over the infrared "fingerprint" region. If we assume 50 gaseous molecular species in the atmosphere, each with 200 reasonably strong lines in this range, the separation between individual lines should average 0.43 cm^{-1}, or about two line-widths. Consequently, even at atmospheric pressure, high specificity should be possible.

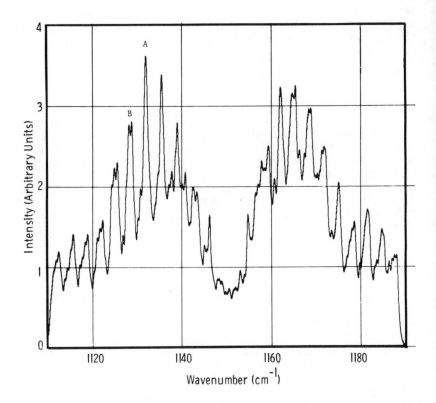

Figure 3.4.   Computed spectrum of 8.7-μm band of SO_2 at 200° C, with 0.3 cm^{-1} full width at half-maximum, based on diode laser scans of this region and microwave data.   See Reference 5 for details.

*Single Wavelength vs
Two-Wavelength Operation*

Figure 3.5 illustrates two modes of operation for tunable diode lasers.   In *a* the laser is tuned to the peak of a strong absorption line at λ_1 of the pollutant gas, whereas in *b* the laser wavelength alternates between λ_1 and λ_2, a region of lesser absorption.   Each method is based on the Beer-Lambert equation, modified as follows:

$$P = P_o \exp(-\alpha'cL) \exp(-\beta L) \tag{1}$$

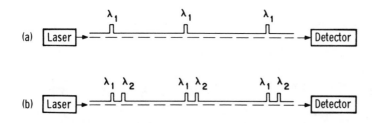

Figure 3.5. Schematic representation of two modes for diode laser pulsed operation: (a) single-wavelength at center of absorption line; (b) two wavelengths alternatively "on" and "off" absorption line.

P_O and P are the transmitted and received laser powers, respectively, α' the gaseous absorption coefficient (cm^{-1}/ppm), c the concentration in ppm, β a scattering or extinction factor that is assumed to vary slowly with wavelength, and L is the total path length. In Figure 3.5(a) a train of pulses of radiation at wavelength λ_1 is emitted. The wavelength may correspond to 1132 cm^{-1} to match peak A in Figure 3.4, for example. As Equation (1) indicates, however, the received laser power is influenced by other factors as well (such as scattering from particulates or atmospheric turbulence) through the parameter β. In order to overcome such broadband "interferences," the two-wavelength technique of Figure 3.5(b) can be used. If the parameter β is independent of wavelength between λ_1 and λ_2, the *ratio* of power P_1 at λ_1 to P_2 at λ_2 will vary only with pollutant concentration. That is,

$$P_1/P_2 = \exp[-(\alpha_1' - \alpha_2')]cL \qquad (2)$$

where α_1' and α_2' are the absorption coefficients at λ_1 and λ_2, respectively. For example, the wavelengths may be adjusted to match the strong SO_2 absorption at A in Figure 3.4, and the valley between A and B. By using a logarithmic converter to analyze the detector signal, Equation (2) reduces to:

$$V_{out} = G(\alpha_1' - \alpha_2') cL \qquad (3)$$

so that the output voltage is directly proportional to the pollutant concentration. G is the voltage gain after logarithmic conversion.

Sensitivity Calculations

Because of the small concentrations of gaseous pollutants generally present in the atmosphere (in the parts-per-billion range away from major sources), long optical paths are needed to achieve a measurable reduction in transmitted laser power. For regional modeling, where the *average* concentration of a pollutant over a large area is important, integrated-path measurements over one or more kilometers are desirable. This type of information is readily obtained by laser techniques. For source monitoring, concentrations are generally in the parts-per-million region, permitting shorter pathlengths to be used effectively. Multiple-reflection cells can be employed for greater sensitivity, if desired.

If we neglect the scattering factor β in Equation (1), and solve for c_{min} (the minimum detectable concentration) for a gas having an absorption coefficient α', we obtain

$$c_{min} = -\ln(P/P_o)_{min}/\alpha'L \simeq (\Delta P/P_o)/\alpha'L \qquad (4)$$

over a total path L. $\Delta P \equiv (P_o - P)_{min}$ is the minimum detectable power, and may be limited by infrared detector noise or by scintillation caused by a turbulent atmospheric path. In the former case, where detector noise may conservatively be assumed to be less than 10^{-9} watts, the concentration detection limits for most gases will be well below 1 ppb over a 2-kilometer path for a laser power P_o of 1 milliwatt. If, on the other hand, sensitivity is limited by external effects, such as turbulence or aerosol scattering, the laser power level can become relatively unimportant, and sensitivity is determined by the ratio $\Delta P/P_o$. In many cases, it should be possibl to detect a 0.3% change in transmitted power.

Table 3.1 lists five important gases, their detection wavelengths, the diode laser compositions that can be used, the measured absorption-line strengths, and the ultimate sensitivity based upon either detector-limited or scintillation-limited measurements. In the former case, sensitivities range from 0.0001 ppb for CO to 0.005 ppb for SO_2. If atmospheric turbulence proves to be the limiting factor for long-path monitoring, use of the two-wavelength technique (with a small time interval between pulses) should eliminate some of the "noise" and yield sensitivities approaching those for detector-limited operation.

Table 3.1

Ambient Air Monitoring Sensitivity for 2-Kilometer Path

				Sensitivity (ppb)	
Gas	$\lambda(\mu m)$	Diode Composition	$\alpha'(cm^{-1}/ppm)$	$\Delta P = 10^{-9}W$	$\Delta P/P_O = 0.3\%$
CO	4.74	$PbS_{0.82}Se_{0.18}$	5×10^{-5}	0.0001	0.3
O_3	4.75	$PbS_{0.83}Se_{0.17}$	0.5×10^{-5}	0.001	3.0
NO	5.31	$PbS_{0.62}Se_{0.38}$	1×10^{-5}	0.0005	1.5
SO_2	8.88	$Pb_{0.93}Sn_{0.07}Te$	0.1×10^{-5}	0.005	15
C_2H_4	10.54	$Pb_{0.88}Sn_{0.12}Te$	2×10^{-5}	0.0003	0.8

EXPERIMENTAL RESULTS

In this section we describe measurements of C_2H_4 over a relatively long atmospheric path, C_2H_4 in the exhaust of an operating automobile, and the detection of SO_2 in the stack effluent of a coal-burning power plant.

Detection of Ethylene Across a Parking Lot

In order to demonstrate the potential of tunable diode laser techniques for long-path, ambient-air measurements, radiation from a semiconductor diode laser was transmitted across a parking lot, as illustrated in Figure 3.6. The laser wavelength was 10.54 μm for the detection of ethylene (C_2H_4), a common hydrocarbon in automobile exhaust. The distance from laser to mirror was 0.25 kilometer, making the total path 0.5 kilometer long. The emitted power P_O was $\sim$ 6 microwatts, and the collection efficiency was around 33%, so that 2 microwatts were measured at the infrared detector. The single wavelength technique of Figure 3.5(a) was used. Figure 3.7 is a recorder scan of the C_2H_4 concentration between 4:51 and 4:56 p.m., when many automobiles were being driven from the parking area, and between 5:05 and 5:10 p.m., when the number was less. With the laser beam approximately two meters above the

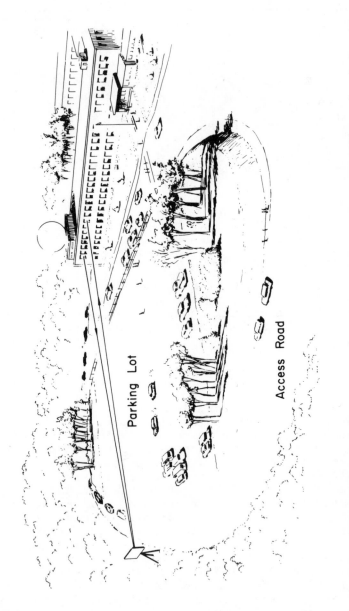

Figure 3.6.   Illustration of experimental configuration for long-path ambient air monitoring. Total pathlength was 0.5 kilometer.

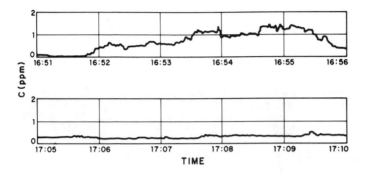

Figure 3.7. *Parking lot measurement of C_2H_4 with $Pb_{0.88}Sn_{0.12}Te$ diode laser during evening rush hour.*

rooftops of the automobiles, the C_2H_4 averaged over the total path is seen to have reached a peak of around 1.3 ppm, diminishing to near zero during the later time interval due to reduced traffic and dispersion. In some cases, C_2H_4 emission from individual automobiles was detected.

Ethylene in Automobile Exhaust

Using the same single-wavelength technique described above, but over a much shorter path, the C_2H_4 content in the exhaust of several automobiles was recorded during a sequence of start-stop cycles. A flexible tube was used to transport the exhaust gases from the tail pipe to a windowless stainless-steel tube (145 cm long) through which the diode laser beam was also directed. Test results for a poorly-tuned 1964 sedan are shown in Figure 3.8. The wide signal fluctuation after ignition is real and is caused by faulty engine operation. The average C_2H_4 concentration is correspondingly high at around 800 ppm. For comparison, Figure 3.9 is a similar presentation from a well-tuned 1972 station wagon having some emissions controls. The wide fluctuations in the signal are absent, indicating a smoother-running engine, and the average C_2H_4 concentration is only 300 ppm. During the second starting sequence of Figure 3.9, the engine was flooded to simulate a heavily-choked condition. During the first four seconds, the recorder pen was driven off the scale twice because of the extremely

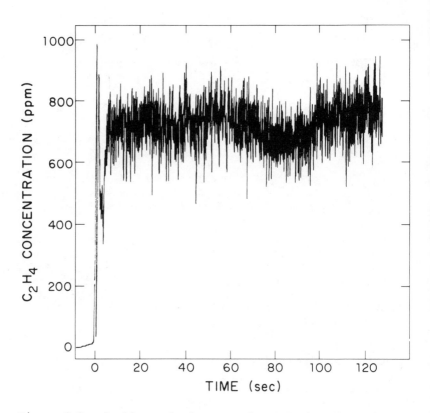

Figure 3.8. *On-line emission test for C_2H_4 in the exhaust of
a 1964 8-cylinder sedan, using a $Pb_{0.88}Sn_{0.12}Te$
diode laser. Ignition occurred at $t = 0$.*

high C_2H_4 content of the exhaust, but the trend after
eight seconds was toward the nominal value for idling
observed in the left-hand trace.

The measurements shown in Figures 3.8 and 3.9
were made using the single-wavelength technique of
Figure 3.5(a). Since the recorded C_2H_4 concentration
from these tests are comparable to "total hydrocarbon"
values obtained by conventional monitoring instruments, one might conclude that the β parameter in
Equation (1) is not negligible. However, we have
made measurements at reduced pressure on samples of
exhaust from these automobiles confirming the C_2H_4
values obtained. Consequently, in this case at
least, scattering and other extinction effects are
small compared to the C_2H_4 absorption.

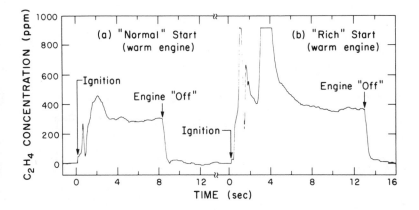

Figure 3.9. *On-line emission test for C_2H_4 in the exhaust of a 1972 station wagon. Two starting cycles illustrated.*

Sulfur Dioxide Detection in a Coal-Burning Power Plant Stack

Measurements of SO_2 concentration and infrared transmission were made at a coal-fired power plant of the Massachusetts Electric Company. The full load capacity was 35 megawatts, produced primarily by bituminous coal at a rate of 16-17 tons per hour, although some oil was being used to reduce particulate emissions. The coal contained 0.8% sulfur, and an Aerotech Cyclone dust precipitator was in service.

The diode laser was positioned at one side of a 5-meter-wide breech between the precipitator and two fans that forced the combustion products into the stack. On the opposite side of the breech was the infrared detector. Another test port was used for the insertion of a Dynascience SO_2/NO_x point-sampling monitor, for comparison.

Figure 3.10 shows the "transmission" and "SO_2 content" of the stack effluent, recorded simultaneously by the laser technique. The 45-second periodicity of the infrared transmission trace is real, although its origin is unknown. The oscillations could be caused by fluctuations in the coal feed rate, or by resonances between the two exhaust fans at the base of the stack. At times, the SO_2 measurements of the Dynascience monitor also exhibited such oscillations, verifying that the condition is due to the stack gas,

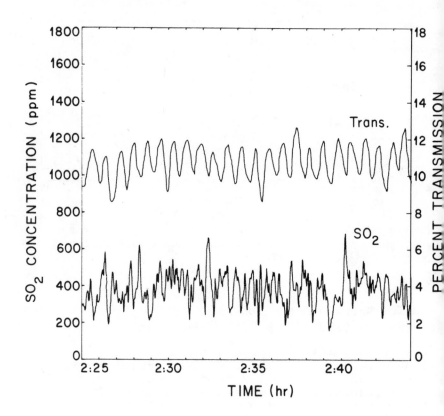

Figure 3.10.   *Transmission and SO_2 measurements obtained by directing radiation from a $Pb_{0.925}Sn_{0.075}Te$ diode laser across the 5-meter-wide stack of a coal-burning power plant. Integration time was 10 seconds.*

and not to an optical alignment change.   The SO_2 signal of Figure 3.10, obtained from the ratio betwee two pulses of slightly different wavelengths (approx mately 1 cm^{-1} apart), in accordance with Figure 3.5(] and Equation (3), shows little of the 45-second periodicity and indicates an average value of 400 ppm. By comparison, the Dynascience point-sampler indicated SO_2 concentrations between 400 and 500 ppm although the measurements were not recorded simultaneously.

ENGINEERING OBSTACLES AND
FUTURE DEVELOPMENTS

The application of semiconductor diode lasers to
problems such as air pollution monitoring has been
limited by both laser fabrication technology and
certain operational requirements. Although the
preparation of semiconductor crystals with appropriate
compositions has been steadily improving, there are
only a few organizations with sufficient skill and
support to provide a sufficient supply of crystals
for those who need them. The same high-level technology is needed for the actual laser fabrication
in order to provide low-resistance electrical contacts
and optimal laser cavity configurations.

At the present time, liquid helium is required
to produce temperatures below 20° K for operation of
the diode lasers described. Pulsed operation has
been achieved at liquid nitrogen temperature (77° K),
and future developments should yield a consistent
supply of diode lasers operable at this temperature
and even higher.

In the interim, tests are being performed with
closed-cycle cryogenic coolers capable of reaching
temperatures as low as 10° K, which require only
electrical power for operation, completely eliminating
the need for liquid helium. Initial experiments with
one commercial model have shown that the temperature
stability has been high enough to permit the recording
of Doppler-limited spectroscopy analogous to that in
Figure 3.3. The closed-cycle systems for diode laser
cooling should then be more than adequate for
monitoring pressure-broadened spectral lines of
atmospheric gases.

CONCLUSION

Tunable semiconductor diode lasers show promise
for future sensitive and specific air pollution
monitoring applications, both in the ambient air and
at sources. Because of their wide wavelength coverage
in the infrared, nearly all of the important pollutant
gases can be detected with these devices, and their
small size is attractive from an instrumental point
of view. Nevertheless, a great deal of research and
development remains to be done to increase the
availability of the diode lasers and to improve their
operating requirements before their utilization can

become widespread. Advances in the production of
the semiconductor crystals and lasers are constantly
being made, and it appears that relatively simple
closed-cycle coolers are able to provide the
necessarily low operating temperatures without the
need for liquid cryogens.

ACKNOWLEDGMENTS

We are pleased to acknowledge the competent technical
support of J. O. Sample, L. B. McCullough, L. B. Belanger, and
W. F. McBride. Appreciation is also expressed to officials of
the Massachusetts Electric Company for assistance in experiments performed at one of their generating stations, and to
Charles E. Rodes of the Environmental Protection Agency for
making the correlative point-sampling stack measurements.
This work was supported by the Environmental Protection
Agency and the National Science Foundation (RANN).

REFERENCES

1. Hinkley, E. D. "Tunable Infrared Lasers and their
    Applications to Air Pollution Measurements," *J. Opto-
    Electronics, 4,* 69 (1972).
2. Hinkley, E. D. and P. L. Kelley. "Detection of Air
    Pollutants with Tunable Diode Lasers," *Science, 171,* 635
    (1971).
3. Calawa, A. R., T. C. Harman, M. Finn, and P. Youtz.
    "Crystal Growth, Annealing, and Diffusion of Lead-Tin
    Chalcogenides,"   *Trans. Met. Soc. Amer. Inst. Mining.
    Eng., 242,* 374 (1968).
4. Hinkley, E. D. "High-Resolution Infrared Spectroscopy
    with a Tunable Diode Laser," *Appl. Phys. Lett., 16,* 351
    (1970).
5. Hinkley, E. D., A. R. Calawa, P. L. Kelley, and S. A.
    Clough. "Tunable-Laser Spectroscopy of the ν_1 Band of
    SO_2," *J. Appl. Phys., 43,* 3222 (1972).
6. Burch, D. E., J. D. Pembrook, and D. A. Gryvnak.
    "Absorption and Emission by SO_2 Between 1050 and 1400
    cm^{-1} (9.5 - 7.1 μm)," Final Technical Report No. U-4947
    of Philco-Ford Corp., July, 1971.

CHAPTER 4

A CHEMILUMINESCENT APPROACH TO
MEASUREMENT OF STRONG ACID AEROSOLS

T. G. Dzubay, H. L. Rook,* and R. K. Stevens

INTRODUCTION

The smog disasters of the Meuse Valley in 1930,
of Donora in 1948, and of London in 1952 have stimulated attempts to develop measuring methods for
aerosols of strong acids in ambient air. A method
developed by Commins, which gained some acceptance,
measures the hydrogen ion concentration,[1] but basic
or weak acid materials are potential interferences.
Scaringelli and Rehme[2] were successful in developing
a sensitive and fairly specific method for sulfuric
acid aerosol. Pretreated fiber filters were used to
collect aerosol in the field and were then returned
to the laboratory. The filters were heated to 400°C,
and the evolved SO_3 was converted to SO_2 by hot
copper and analyzed. Although ammonium sulfate would
be detected as an interference by this method, that
compound is nearly comparable to H_2SO_4 as a pulmonary
irritant.[3] More recently Barton and McAdie have
developed an automatic field monitor for H_2SO_4.
Aerosol is collected on a polycarbonate filter tape
and is eluted with isopropanol.[4] This is followed
by colorimetric analysis with barium chloranilate.
In the above methods there may be some risk of
loss of the strong acid due to reactions taking place
on the required collection surface, although the use
of pretreated surfaces and short collection times

---

*On interagency transfer from the National Bureau of Standards.

reduces that risk.[2,4]   The present work represents
an effort to develop an instrument for monitoring
strong acids in real time without going through the
collection and extraction operations.

DESCRIPTION OF METHOD

Strong acid aerosol is detected by its gas phase
reaction with ammonia, which is supplied at a constant
rate by a permeation tube.   In the system illustrated
in Figure 4.1, the resulting ammonium salts are removed by filter F_2, and the remaining NH_3 is converted
to NO in a heated converter.   The NO is detected with

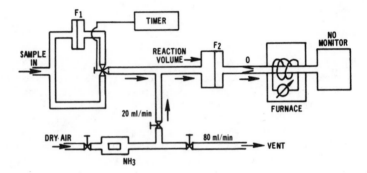

Figure 4.1.   Schematic view of strong acid mist monitor.

the chemiluminescence monitor illustrated in Figure
4.2.[5,6]   In this monitor NO reacts with O_3 from an
ozone generator to produce photons according to:

$$NO + O_3 \rightarrow NO_2^* + O_2 \tag{1}$$

$$NO_2^* \rightarrow NO_2 + h\nu \tag{2}$$

The photons, which pass through a 600 nm cutoff filter
are detected by a photomultiplier tube for which the
output is proportional to the NO concentration.
With the above configuration an increase in the
strong acid concentration results in a proportional
decrease in the output signal.   However, ambient

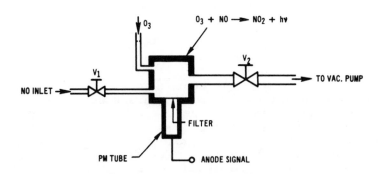

Figure 4.2. *Schematic view of chemiluminescence monitor for*
NO. Valves V_1 and V_2 are used to adjust the inlet
flow resistance and outlet pumping speed
respectively.

levels of NO and NO_2 may significantly contribute to
this signal as an interference. The interference is
eliminated by periodically passing the sample air
through the filter F_1 (shown in Figure 4.1), which
removes aerosol. Therefore, the instrument response
as a function of time is a square wave with amplitude
proportional to the strong acid concentration.

There are a variety of phenomena that can affect
the accuracy of this method. Each must be considered,
and its effects must be minimized. For a commercial
chemiluminescence monitor the typical sensitivity is
5 ppb for NO. This corresponds to a molar equivalent
of 8 $\mu g/m^3$ of SO_3. In order to realize this sensitivity, one must be concerned that the periodic flow
resistance changes caused by filter F_1 of Figure 4.1
do not produce an erroneous response. In addition,
one must show that the conversion of NH_3 to NO is
quantitative and that the reaction of the acid aerosol
with ammonia is complete. Effects due to humidity
changes, to reactions with weak acids, and to wall
losses must be minimized.

EXPERIMENTAL INVESTIGATIONS

Stability of Chemiluminescence
Monitor Against Flow Changes

In the chemiluminescence monitor shown in Figure
4.2, needle valves V_1 and V_2 were used to control
the flow resistance at the chamber inlet and outlet,
respectively. A change in V_1 can simulate a flow
resistance change in the acid scrubbing filter F_1
of Figure 4.1. A change in V_2 can simulate a change
in the pumping speed of the vacuum pump. It was
found that small changes in either V_1 or V_2 caused
serious changes in the instrument response. One can
completely evaluate the stability of the instrument
by studying the effects due to V_1 and V_2.
A series of measurements were made with various
settings of pump speed valve V_2 causing the reaction
chamber pressure to range from 6.5 to 560 Torr. For
each measurement the inlet valve V_1 was adjusted to
maintain a flow rate of 165 $\pm$ 5cm^3/min. This sample
consisted of dry air containing 44 ppm of NO flowing
at a rate of 12.0 $\pm$ 0.2 cm^3/min diluted with enough
additional dry air to satisfy the 165 cm^3/min demand
at the inlet orifice.
The results are shown in Figure 4.3. The maximum
sensitivity occurs at a pressure of about 25 Torr.
The sensitivity decreases at higher pressures because
of increased quenching of the excited NO_2^* species
described in Equation 1.[5] When the pressure is
decreased below 25 Torr the sensitivity decreases
because the residence time of the molecules in the
reaction chamber becomes small compared to the mean
time required for NO to react with O_3 as in Equation
1. An expression based on the reaction kinetics
and flow conditions is also plotted in Figure 4.3
and is found to be in reasonable agreement with the
data. For the present flow rates it is concluded
that the greatest sensitivity and stability against
changes in pumping speed occur at an operating
pressure of 25 Torr.
The response to changes in the input valve V_1 of
Figure 4.2 was also investigated. For each measurement the initial flow rate was set at 165 cm^3/min,
and the initial pressure was set at a value in the
10 to 500 Torr range of investigation. Then the
inlet valve V_1 was opened slightly, which caused
about a 15 per cent increase in both flow rate and
pressure. The change in output signal is shown as

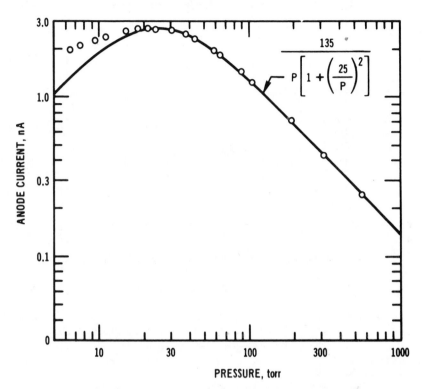

$$\frac{135}{P\left[1 + \left(\dfrac{25}{P}\right)^2\right]}$$

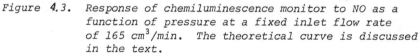

Figure 4.3. *Response of chemiluminescence monitor to NO as a function of pressure at a fixed inlet flow rate of 165 cm³/min. The theoretical curve is discussed in the text.*

a function of operating pressure in Figure 4.4. The signal decrease associated with an increase in input flow rate is due to the corresponding increase in reaction chamber pressure and hence in quenching. A theoretical expression derived from the known flow conditions is also plotted in Figure 4.4 as a horizontal line and is in good agreement with the data.

From Figure 4.4 one concludes that the inlet flow resistance changes, which are associated with the periodic introduction of the acid scrubbing filter F_1 of Figure 4.1, can lead to a serious error signal. This error can be minimized by using a glass

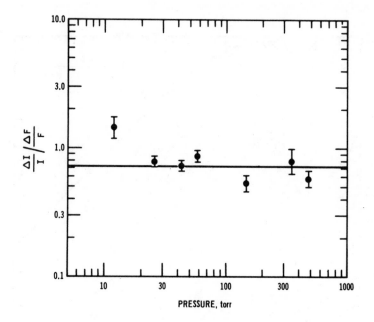

Figure 4.4. Response of the chemiluminescence monitor to changes
in inlet flow resistance. The straight line is
discussed in the text.

filter for F_1, which has a high collection efficiency
and a low flow resistance.[7]

Measurements with Stainless Steel and Copper Converters

It was necessary to establish quantitative conversion of NH_3 to NO. As an initial effort a stainless steel tube of 1-cm inside diameter was tested
as a function of temperature in a 30-cm long furnace.
Dry air containing 600 $\mu g/m^3$ of ammonia was drawn
through the converter at a rate of 280 ml/min. The
response of the chemiluminescence monitor operating
at a pressure of 200 Torr is plotted in Figure 4.5.
The conversion efficiency was not quantitative but
increased exponentially as the temperature was raised
from 600 to 800°C. In a previous study a plateau
was observed for temperatures above 500°C,[6] but
presumably the nature and history of the stainless
steel tube can be of critical importance.

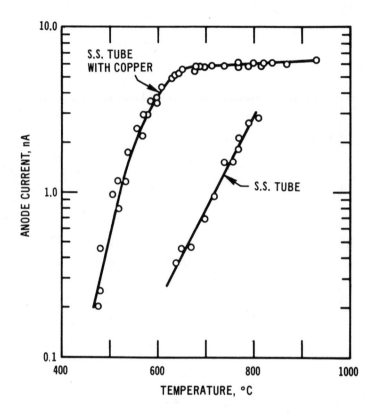

Figure 4.5.   *Conversion of NH₃ to NO as a function of tempera-*
ture for a stainless steel tube and a copper-packed
stainless steel tube.

A great improvement in conversion efficiency was
obtained by loosely packing the stainless steel tube
with copper wool, which becomes oxidized. In the
conversion curve plotted in Figure 4.5, there is a
broad plateau for temperatures between 700 and 900°C,
where the conversion efficiency is assumed to be
100%.
Because the filters in the system could con-
ceivably alter the moisture content of the air being
sampled, the effects of changing the relative humidity
were studied. A stream of clean air flowing at
160 cm³/min was set up so its relative humidity could
be quickly changed from 10% to 90% by bubbling it

through water.   This stream was mixed with air flowing at a rate of 20 cm^3/min containing either NO or
ammonia.   The change in signal due to the above
change in humidity is plotted in Figure 4.6.   When

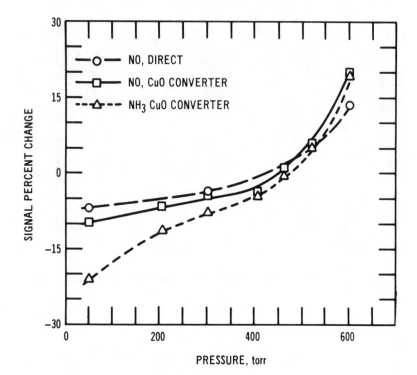

Figure 4.6.   *Sensitivity to changes in signal output as a
function of pressure when the relative humidity is
changed from 10 to 90%. Results are shown for NO
bypassing the converter and for NO and NH$_3$ passing
through the copper-packed stainless steel tube
operated at 800°C.*

the chemiluminescence reaction chamber is operated
at a pressure of about 450 Torr, the effect due to
a change in relative humidity is small. However,
this effect is quite significant at lower and at
higher pressures.
      Attempts to operate this system as a monitor for
sulfuric acid aerosol were unsuccessful when the
copper-packed stainless steel tube was used as a

converter. The chemiluminescence reaction chamber was operated at pressures of either 25 Torr or 460 Torr. Rather than observing the expected square wave response to an acid aerosol, large nonreproducible transients were observed. It was concluded that the copper converter was the source of the problem.

Results with Gold Converter

The difficulties encountered using copper in the converter tube were overcome by using gold. Figure 4.7 shows a conversion efficiency curve for ammonia as a function of temperature of a 30-cm long oven for a 0.4-cm I.D. stainless steel tube packed with gold wool. The conversion of NH_3 to NO is almost

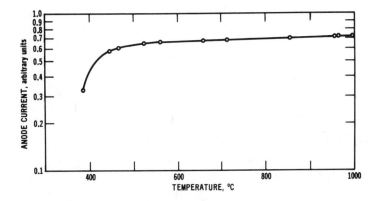

Figure 4.7. Conversion of NH_3 to NO in a stainless steel tube packed with gold wool.

constant and assumed to be complete for temperatures between 600 and 1000°C. A temperature of 1000°C was selected as the operating point for the remainder of this work.

The system was set up as in Figure 4.1. A 47-mm diameter glass fiber filter was mounted in a Lexan plastic holder and used as filter F_1 for periodic acid aerosol removal. A 47-mm diameter membrane mounted in a stainless steel holder was used as F_2 for removal of ammonium salts. Teflon tubing with an inside diameter of 0.7 cm was used to connect filters F_1 and F_2 with the three-way valve. Filter

F_2 and the tubing containing a mixture of NH_3 and aerosol were operated at a temperature of 50°C in order to minimize the response time. The sampling cycle had a period of 18 minutes and was adjusted to pass the sample through F_1 for 16 minutes and to pass it directly into the system for 2 minutes. The chemiluminescence reaction chamber was operated at a pressure of 25 Torr.

The permeation tube was operated at a temperature of 8°C where its output was 0.2 µg/min. Only one-fifth of this output was carried to the acid-ammonia reaction volume into which the total flow was 0.2 l/min giving an ammonia concentration of 200 µg/m³. This should be sufficient ammonia for complete reaction with almost 600 µg/m³ of H_2SO_4 aerosol.

Aerosols of H_2SO_4 were generated in the 100 to 2000 µg/m³ range of concentrations utilizing the same generator used by Scaringelli and Rehme.[2] Dry air flowing at rates between 10 and 100 cm³/min was passed over fuming sulfuric acid and then diluted with air from the room flowing at 2 to 5 l/min. This mixture passed through a 40-liter carboy vessel where the aerosol was allowed to age and grow in particle size.

The behavior of the complete strong acid monitor is shown in Figure 4.8. When the sample inlet was simply held out in the room, the response was essentially constant as a function of time. When the sample inlet was inserted into the carboy vessel, the expected characteristic acid response was observed. When the sample line was withdrawn from the acid atmosphere, the signal returned to an equilibrium value after about 30 minutes. Figure 4.9 shows the response to H_2SO_4 aerosol over a longer period of time. The amplitude of the waveform with the 18 minute period (which indicates the acid concentration) is reasonably constant even though there is considerable variation in the baseline, which is presumed to be due to variations in the NO and NO_2 levels in the room.

The H_2SO_4 concentration for the data of Figures 4.8 and 4.9 was determined by collecting the aerosol on a 0.45-µm pore size membrane filter and analyzing for elemental sulfur with an energy dispersive x-ray fluorescence spectrometer.[8] By analyzing both sides of the filter it was concluded that the acid aerosol was being collected on the front surface, making it unnecessary to correct the x-ray results for attenuation in the filter medium. Using this procedure the concentration was measured to be 200 ± 100 µg/m³.

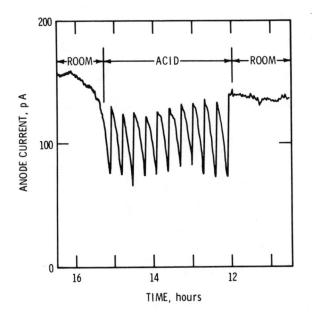

Figure 4.8. *Response of strong acid monitor to clean room air and to 200 µg/m³ of H₂SO₄ aerosol.*

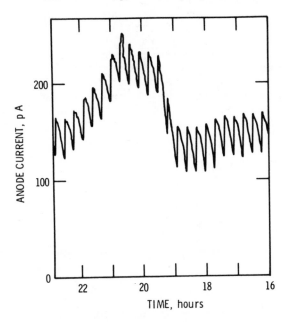

Figure 4.9. *Response to 200 µg/m³ of H₂SO₄ aerosol suspended in air having a slowly varying NO and NO₂ concentration.*

The accuracy figure reflects the possibility of
sampling errors due to possible concentration gradient
within the carboy vessel.

DISCUSSION

Figures 4.8 and 4.9 show that the instrument
gives a clear and unique response to H_2SO_4 aerosol,
and that slowly varying background levels of NO and
NO_2 can be tolerated. From the magnitude of the
peaks in Figures 4.8 and 4.9 it can be concluded
that the detection limit is about 20 $\mu g/m^3$ for sulfuric acid aerosol. The sensitivity of this method
for measuring strong acid aerosols will improve as
more sensitive NO monitors are developed. Improved
design techniques that minimize the retention time
of ammonia on the walls of the system would allow
one to shorten the period of operation. This would
permit signal averaging over several periods in order
to realize improved sensitivity and accuracy.

The question of interferences by weak acid gases
has not been specifically investigated. However,
the interference can be eliminated by choosing the
flow rate and reaction volume (see Figure 4.1) so
that the residence time is smaller than the reaction
time. The residence time is sufficiently short so
that CO_2 in the atmosphere is not an interference as
is illustrated by Figure 4.8 when room air is being
sampled. On the other hand, further work is needed
to test other weak acid gases and also to show that
the residence time is sufficiently long to insure
that the acid aerosol undergoes a complete reaction.

ACKNOWLEDGMENT

The authors are grateful to J. A. Hodgeson and K. A. Rehme
for very valuable discussions and suggestions related to this
project.

REFERENCES

1. Commins, B. T. "Determination of Particulate Acid in
    Small Town Air," *Analyst, 88,* 364 (1963).

2. Scaringelli, F. P. and K. A. Rehme. "Determination of Atmospheric Concentrations of Sulfuric Acid Aerosol by Spectrophotometry, Coulometry, and Flame Photometry," *Anal. Chem.*, *41*, 707 (1969).

3. Amdur, M. O. "The Impact of Air Pollutants on Physiologic Responses of the Respiratory Tract," *Proc. Am. Philosophical Soc.*, *14*, 3 (1970).

4. Barton, S. C. and H. G. McAdie. "A Specific Method for the Automatic Determination of Ambient H_2SO_4 Aerosol," in *Proceedings of the Second International Clean Air Congress*, H. M. Englund and W. T. Beery, Eds. (New York: Academic Press, 1971), p. 379.

5. Fontijn, A., A. J. Sabadell, and R. J. Ronco. "Homogeneous Chemiluminescent Measurement of Nitric Oxide with Ozone," *Anal. Chem.*, *42*, 575 (1970).

6. Hodgeson, J. A., J. P. Bell, K. A. Rehme, K. J. Krost, and R. K. Stevens. "Application of a Chemiluminescence Detector for the Measurement of Total Oxides of Nitrogen and Ammonia in the Atmosphere," in Joint Conference on Sensing of Environmental Pollutants, Paper No. 71-1067, American Institute of Aeronautics and Astronautics, New York (1971).

7. Lockhart, L. B., Jr., R. L. Patterson, Jr. and W. L. Anderson. "Characteristics of Air Filter Media Used for Monitoring Airborne Radioactivity," NRL Report 6054 (1963).

8. Goulding, F. S. and J. M. Jaklevic. "X-Ray Spectrometer for Airborne Particulate Monitoring," EPA Report, EPA-R2-73-182. National Environmental Research Center, Research Triangle Park, N.C. (1973).

CHAPTER 5

A NEW TECHNIQUE TO MEASURE SULFURIC ACID
IN THE ATMOSPHERE

P. S. Mudgett, L. W. Richards, and J. R. Roehrig

It has long been known that air pollution can be
harmful to human health. When only a few atmospheric
constituents are monitored to obtain information about
the air quality, SO_2 is usually included among them,
so more data are available for this pollutant than
for most others. As a result, SO_2 levels are sometimes used as an indicator of the general level of
air quality. A recent analysis of the daily rate
of mortality from all causes in a major metropolitan
area has shown that this rate correlates with the
SO_2 level over the entire concentration range observed
there, provided other more important predictors of
this rate are properly taken into account.[1] However,
studies with monkeys exposed to high concentrations
of SO_2 for two years, and with other animals exposed
to high concentrations of various sulfur compounds
and combinations of compounds for shorter periods of
time, suggest that SO_2 itself may not be dominant in
causing the health effects associated with high SO_2
levels. There is the possibility that sulfuric acid
and other sulfates formed from the oxidation of SO_2
in the atmosphere may be largely responsible for
these effects. This idea has generated a great
interest in these compounds.
    In order to determine the effect of various sulfur
compounds on human health, it will be necessary to
monitor their concentrations in air. At the present
time, no instrument is commercially available for the
routine monitoring of either total sulfates or sulfuric acid, but an instrument for monitoring sulfuric
acid has been described.[2] Forrest and Newman have

85

reviewed the previous work on monitoring various sulfur compounds in the atmosphere.[3]

The work described here has the goal of developing a prototype instrument for the routine monitoring of sulfuric acid in the ambient air. The method selected for development has been described by Richards[4] and has the following steps:

1. The ambient air is passed through a filter to collect the sulfuric acid (as well as the other particulates present).
2. The sulfuric acid is separated from the other particulates by volatilizing it with warm dry air.
3. The amount of sulfuric acid volatilized is determined with a commercially available flame photometric detector (FPD), which is highly sensitive and specific to sulfur-containing compounds.

This method of analysis is related to one reported by Scaringelli and Rehme[5] except that it places a new emphasis on dryness rather than heat for the volatilization. This has the advantage of eliminating the interference from ammonium sulfate and reducing the possibility of other interferences. Dubois and co-workers have also reported a manual method of analyzing filters for sulfuric acid, which involves volatilization, but it does not lend itself to mechanization.[6,7] In the monitoring method proposed here, the serial steps of filtration and volatilization will separate sulfur-containing compounds so that there should be no important interferences. It is expected that the major interference will be the reaction of sulfuric acid with other particulates on the filter. The sensitivity of the method seems to be controlled by the ability to volatilize quantitatively fractions of a microgram of sulfuric acid and transport it to the FPD, rather than the inherent sensitivity of the FPD itself.

The majority of this chapter is a report on some preliminary experiments that have been done to see if the proposed monitoring method could be reduced to practice. The experiments were successful enough that a prototype instrument has been built to carry out the operations automatically. At the time of writing, only enough experiments to show the validity of the preliminary experimental results reported here have been done with the prototype.

PRELIMINARY EXPERIMENTS

The preliminary experiments were motivated by
the goal of providing the information necessary to
design the prototype instrument rather than developing
a set of data to demonstrate the capabilities of a
new monitoring method. Therefore the results presented here are somewhat limited. The prototype
monitor, which operates automatically and gives much
more reproducible results than were obtained from
these manual experiments, will generate the data on
which the evaluation of the technique can be based.
Aside from the flame photometric detector, most of
the equipment and materials used for these experiments
were selected from those already on hand in the
laboratory.

Flame Photometric Detector
Operation and Calibration

A Meloy Laboratories model 100AT flame photometric
detector (FPD) purchased from TRACOR, Inc. was used.
It has an aluminum flame housing (not Teflon-coated)
and a 393-nm narrow band pass interference filter.
The bias for the photomultiplier was obtained from
a Fluke 410B regulated high voltage power supply.
A Keithley 610BR electrometer amplifier with a
Moseley Autograf Model 680 strip chart recorder was
used to record the output. The high voltage to the
photomultiplier was -750 V. The typical dark current
was 1 to 2 na, and the typical flame on, zero sulfur
signal,was 10 na.
The hydrogen and air for the flame were obtained
from commercial cylinders; they were Air Reduction
Co. prepurified (99.95%) grade and Linde dry grade,
respectively. The air was stated to have a moisture
content not exceeding 3 ppm, and was passed through
P_2O_5 to ensure its dryness. The gas flow rates
usually used were 200 std cc/min of air and 150 to
175 std cc/min of hydrogen. The FPD burner block
was maintained at a temperature of 210-230°C, with
the exception of temperatures near 170°C used in some
early experiments.
In order to interpret the signals from the FPD,
it was necessary to determine the relative response
to sulfuric acid vapor. This was done by mixing
known flow rates of dry air and air containing sulfuric acid vapor. The output signal of the FPD
depends on the air flow rate, so the flow system was

designed in such a way that the total flow rate of
air was kept constant when the fraction of flow containing sulfuric acid was varied. The results are
shown in Figure 5.1, where I is the FPD output current
and I_O is the output current in the absence of sulfur.
Since the absolute concentration of the sulfuric acid
is unknown, this is a relative calibration, and only
the slope of the lines in Figure 5.1 is of importance.
The average value of the slope is 1.6, which is
smaller than usually observed for other sulfur-
containing compounds. Once this slope has been
determined, it becomes possible to integrate FPD
signals which vary with time to obtain a number
proportional to the total amount of sulfuric acid
detected.

Sulfuric Acid Aerosol Generator

The aerosol generator has the design reported
by LaMer, Inn and Wilson.[8] The information given
in that article is adequate to describe the generator
used in this work.

Apparatus and Procedures for the Volatilization Experiments

The gas flow system used in the volatilization
experiments is shown in Figure 5.2. The hydrogen
and air flow meters were Brooks 150-mm variable area
low flow metering tubes. About 20 cc/min of air
always passed through valve C directly to the FPD
so that the flame would not go out when valve A was
turned. Valve A permits the majority of the air
flow to go either directly to the flame so that the
filter holder can be opened or through the filter
from which the sulfuric acid is being volatilized.
Since the air flow is controlled and metered ahead
of this valve, the zero sulfur signal does not depend
on the valve position. Therefore the zero sulfur
signal can be obtained at any time during a volatilization run by operating valve A so that all the air
flow goes through the bypass line.
The filter holder has been made from a variety
of materials, but aluminum or Teflon gave the best
performance. The heater on the air inlet and the
thermocouple on the heated tube permitted the air
temperature to be rapidly varied and also monitored.
The flow system between the filter and the FPD was
always heated, and was usually maintained between
170 and 180°C.

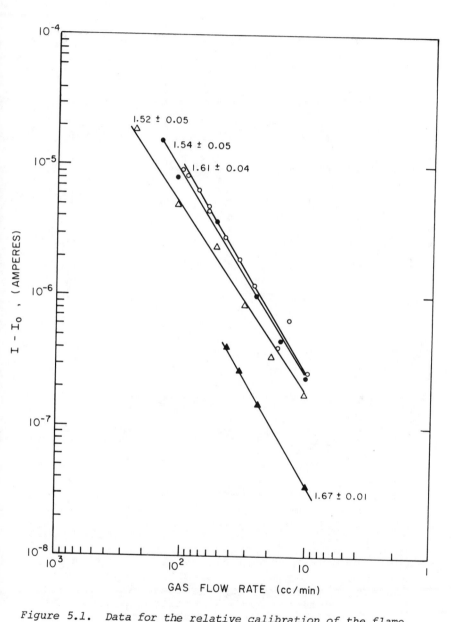

Figure 5.1. Data for the relative calibration of the flame photometric detector. The concentration of sulfuric acid reaching the flame is proportional to the gas flow rate. The numbers beside each line give the slope and the standard error of the slope as determined by a regression analysis.

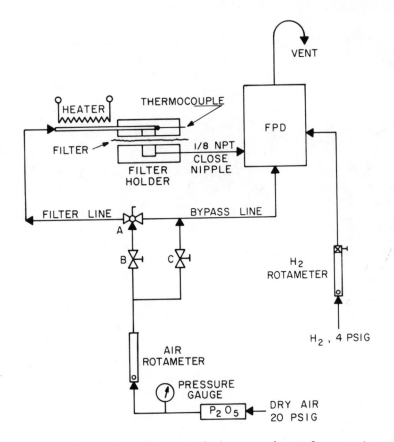

Figure 5.2.   Gas flow diagram of the experimental apparatus.

The very first experiments were done at lower temperatures because the vapor pressure tables for sulfuric acid indicate that acid in equilibrium with dry cylinder air at about 50°C is volatile enough to overload the FPD. A record from one of these early runs is shown in Figure 5.3. Unfortunately, subsequent experiments showed that appreciable amounts of acid were not volatilized even though the FPD signal returned almost to the base line. Therefore higher temperatures are now always used.

A typical procedure for a volatilization experiment is as follows:

1. The FPD is left running with room temperature, dry air flowing through the filter holder unit

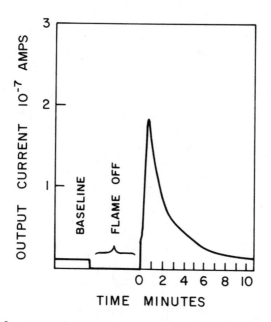

Figure 5.3. *Record of the volatilization of an aerosol sample at a temperature of 44°C. The transfer line to the FPD was maintained at 100°C.*

while the filter containing sulfuric acid is being prepared.

2. The air is switched to the bypass line, the filter placed in the holder, and the air switched to flow through the filter.

3. The temperature of the dry air flow is manually controlled by adjusting the heater power according to the amount of sulfuric acid on the filter and the response of the FPD.

   a. With very little sulfuric acid, the heater power is increased rapidly at the first sign of the FPD response to sulfuric acid.

   b. With about 1 µg sulfuric acid, the heater power is gradually increased until the FPD gives a maximum response, and then is set at the maximum power.

   c. With two or more micrograms of sulfuric acid, the heater power is very slowly increased until a FPD response is obtained, and then

is adjusted to maintain the desired response.
Near the end of the run the response falls
off rapidly even though the heater power is
increased to the maximum. Figure 5.4 is an
example of a record from such a run.

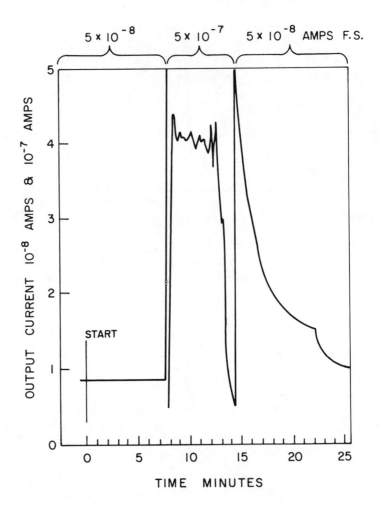

Figure 5.4.   *A sample record of the FPD output during the vola-*
              tilization of a single drop of dilute sulfuric aci
              containing 2.5 μg H_2SO_4. The dry air temperature
              was 100°C at the start of the volatilization, 130°
              at the end of the peak, and 220°C when the heat wa
              turned off at a time of 23 min. The temperature
              was manually controlled to keep the output current
              near 4 x 10^{-7} amps.

4. The volatilization is continued with the heater at maximum power until the FPD signal has decreased to a predetermined value, at which point the heater is turned off. The air temperature is usually 220-240°C and rising slowly at this point in the run. It begins to fall rapidly when the power is turned off.

5. The run is terminated 1 to 3 min after the heater power is turned off by turning the air flow to bypass and removing the filter.

At the start of each day, and sometimes more frequently, the above procedure was followed with no filter in the holder to show that the system was free of sulfur compounds. Each time a new filter was used, and sometimes more frequently, the procedure was followed without adding sulfuric acid to the filter to see if there were volatile sulfur compounds on the filter. After some experiments with large amounts of sulfuric acid (above 5 µg), it was necessary to wash out the apparatus if the next experiments were to be done with very small amounts of acid.

Filter Materials

The performance of a number of kinds of filter materials was explored early in the work. These early experiments are less reliable than the later ones, but we believe that the conclusions drawn from them are accurate enough that there is no need to repeat the survey of filter materials at this time. The results can be summarized as follows. Only Teflon gave satisfactory performance. Glass fiber filters retained much more sulfuric acid than Teflon, and other materials reacted with the acid to produce discoloration and sometimes disintegration of the filter.

Many of the early experiments, including those reported below, were performed with Millipore Mitex filters because they contain only Teflon. The smallest pore size available is 5 µm. These filters were convenient to use in the early experimental work, but have a larger pore size than is desirable when monitoring ambient air. The prototype monitor uses Millipore Fluoropore filters with a 0.5-µm pore size that have been treated with fuming sulfuric acid to remove the polyethylene backing.

Early Experimental Results

The first experimental results, such as the recor
in Figure 5.3, were deceptively promising. Attempts
to extend the experiments to the smallest possible
amounts of acid showed that it is difficult to quantitatively volatilize the sulfuric acid from the
filter. It now appears that the volatilization of
the sulfuric acid is never entirely complete; this
has been verified by some radioactive tracer experiments described below. Therefore the goal of the
preliminary experiments was to find conditions where
the volatilization was nearly complete and was
reproducible, so that the instrument could be
calibrated.

The three methods were used to apply sulfuric acid
to the filter. In some experiments, a known volume
of sulfuric acid aerosol was drawn through the filter
before it was placed in the volatilization apparatus.
The amount of acid on the filter was determined by
drawing larger volumes of the same aerosol through
a similar filter, washing off the collected acid
with methanol, adding water, and measuring the
electrolytic conductance.

The simplest method of applying acid was to use
a microsyringe to place between 5 and 10 μl of a
dilute aqueous solution of sulfuric acid on the
filter. This remains as a single drop during the
volatilization, so that no sulfuric acid vaporizes
until the drop dries. Figure 5.4 is an example of
such a run, and it can be seen that the onset of
volatilization of acid is quite sudden. This is an
indication that adequate dryness is critical. It is
also possible to apply solutions of sulfuric acid in
methanol to the filters. Methanol can wet the
Teflon filter material, permitting the sulfuric acid
to be spread over the filter so it forms many small
droplets when the methanol dries. Thus a microsyrin
can be used to dispense a known amount of acid in a
manner that simulates the acid deposit from an aeros

The experimental results are reported as a relative integrated signal (RIS), which is the integral
of the linearized flame photometric detector respons
divided by the amount of sulfuric acid used in the
experiment:

$$RIS = \frac{10^4 \int (I - I_o)^{1/1.6} \, dt}{\mu g \ H_2 SO_4}$$

Here I is the FPD output current in amperes, I_O is the zero sulfur signal, the exponent 1/1.6 was taken from the relative calibration data, and time t is measured in minutes.

For the larger amounts of acid, the FPD response is proportional to the amount of acid taken. The early experiments included only eight runs using 2.4-17 μg acid. These experiments have a mean RIS of 2.7 and a standard deviation of 0.55. For amounts of acid smaller than 2 μg, the RIS becomes smaller, as if a fraction of a microgram of acid always fails to reach the FPD. For example, a series of experiments with 0.54 μg of sulfuric acid applied in a single drop of water gave a mean RIS of 1.1_5 and a standard deviation of 0.2. Three similar experiments performed a month later gave a mean RIS of 1.1_5 and a standard deviation of 0.3. Results of this kind show that the completeness of the volatilization is reproducible. The prototype monitor gives smaller standard errors than were obtained from these early experiments.

The Use of Volatilization
Aids (Methanol)

It was discovered in the course of this work that the addition of about 200 ppm of methanol to the dry air flow increased the volatilization of the sulfuric acid from the filter and reduced the adsorption of sulfuric acid between the filter and the flame. It is surmised that this effect is due to competitive adsorption, and that the methanol tends to displace the adsorbed sulfuric acid. However, it was not thought appropriate to do much research to develop an understanding of this observation during the phase of the work reported here. By adding methanol to the hydrogen flow it was shown that the apparent effect of methanol injection is not due to some interaction in the FPD that increases the response of the detector.

The primary benefit of methanol is that the relative integrated signal (RIS) remains constant over a wider range of amounts of acid. Without methanol addition, the RIS begins to fall off when less than 1.5-2 μg of sulfuric acid is applied to the filter. With methanol, the RIS remains constant if more than 0.3 or 0.4 μg of acid is used. For example, nine experiments with 0.57 μg of sulfuric acid gave a mean RIS of 2.8 and a standard deviation of 0.85, which compares well with the RIS reported above for amounts of acid between 2.4 and 17 µg.

Radioactive Tracer Experiments

Experiments with sulfuric acid containing S^{35} provide a convenient means for determining what happens to the sulfuric acid placed on the filter at the beginning of the volatilization. Since a Tracerlab automatic counter with coincidence circuits to reduce the background counting rate was available, it was possible to work with very low levels of radiation, and hence to do these experiments rather simply.

The sulfur 35 isotope has a half life of 88 days, and decays by emitting a beta particle with a maximum energy of 0.167 Mev. Most of the emitted particles have an appreciably lower energy, but no effort was made to learn the energy distribution of the emitted particles. Beta rays with an energy of 0.167 Mev have a penetration of 34 mg/cm^2. The Teflon filters have a thickness of about 7 mg/cm^2, and transmit about 10% of the radiation from S^{35}.

In a series of experiments, 1 µg of sulfuric acid in a few milligrams of methanol was placed on one surface of the filter, and this penetrated the pores to some extent. The acid was doped to have an activity of roughly 20,000 disintegrations per minute, but since not all disintegrations cause a count, the initial activity was in the neighborhood of 4500 cpm for the surface of the filter to which the acid was applied, and 1600 cpm on the other surface. These count rates were higher if there was already some activity on the filter from a previous experiment. It was usually found that 65-80% of the added activity was lost from the filter during a volatilization run, which is consistent with the data given earlier. No methanol was added to the dry air during these volatilizations.

Some of these experiments had the double purpose of observing the adsorption of sulfuric acid on surfaces. Platinum, gold, aluminum and both FEP and TFE Teflon were placed in the gas flow that carries the volatilized sulfuric acid from the filter to the FPD, and the activity picked up by these surfaces was determined. Teflon was better than any of the other materials, and TFE Teflon picked up somewhat less activity than did FEP Teflon. It appears that Teflon is a satisfactory material to carry the

volatilized acid from the filter to the FPD. Tubes
as long as four feet have been used without apparent
degradation of the FPD signal.

These experiments with S^{35}-doped sulfuric acid
have shown that it is unlikely that all the sulfuric
acid on the filter can be volatilized and transferred
to the FPD. However, it is believed that the efficiency of the transfer is reproducible enough that
it can be included in the calibration of the prototype monitor. Probably the most important result
caused by this residual sulfuric acid is a limitation
of the sensitivity of the instrument.

THE PROTOTYPE MONITOR

At the time of writing, the construction of the
prototype monitor was mostly complete, and only a
small amount of laboratory test data had been obtained
from it. The monitor is described in three parts:
gas flow system, filter advance mechanism, and
electronics.

Gas Flow System

The gas flow system is shown schematically in
Figure 5.5. The ambient air enters through a short
length of 1-inch tubing, and immediately passes
through the filter F2, which collects the particulates.
The air then flows to a system for measuring and controlling the air sampling rate, and to a pump where
it is exhausted to the atmosphere. An ambient air
preheater is supplied so that the instrument can be
operated in a fog without collecting excessive
moisture on the filter.

A separate air flow system carries the dry air
that is used to volatilize the sulfuric acid from
the filter and to supply air to the flame photometric
detector. A mechanical system described below cycles
the filters between the two air flows. The dry air
is supplied by a cylinder with a pressure regulator.
It is filtered, reduced in pressure to 20 psig, and
passed through the shut-off valve FC1. The flow
meter-controller FI2 supplies air to the ambient
preheater mentioned above.

The volatilization air flow passes through a
flow indicator and past a methanol injector; it is
then split into two streams. About 10% of this flow

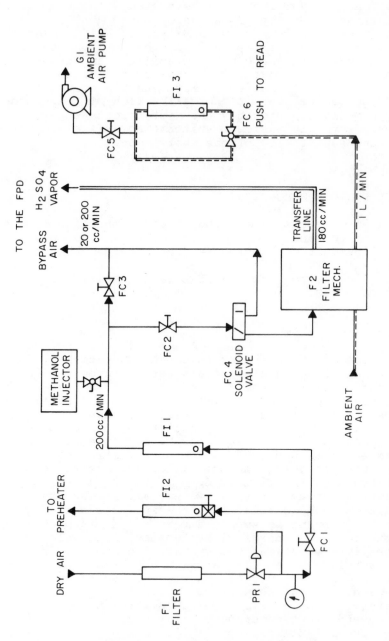

Figure 5.5.   Gas flow system. Some typical flow rates are indicated.

goes through valve FC3 directly to the FPD. This
bypass air tends to even out the air flow to the FPD
during the flow switching associated with changing
the filters, and greatly reduces the chance that the
flame will go out. Except when the filter is being
changed, the remaining air flow goes through valve
FC4, the filter, and then to the FPD. When the
filter holder is open during the filter changing
operation, valve FC4 diverts the air directly to the
FPD.

The flow system is designed in such a way that
the flame-on, zero sulfur signal from the FPD is
easily checked. The rate of flow of air to the FPD
is controlled by valves FC2 and FC3, and is independent of the position of valve FC4. Thus if no
sulfur-containing compounds are added to or subtracted
from the air on passing through the filter unit and
transfer line, the FPD signal will be the same for
both positions of valve FC4. If the drift of the
zero signal from the FPD should turn out to be a
major problem, it will be possible to automatically
recalibrate this zero each cycle while valve FC4 is
in the bypass position. It is not expected that
this will be necessary.

The flame photometric detector is a Meloy Model
SA 160-2, which normally has two air inlets. The
bypass air flow is connected to the inlet that passes
through the flow meter, and the flow containing the
sulfuric acid is connected to the line that goes
directly to the burner. The direct line was modified
so that it is heated.

The air passing through the filter to volatilize
the sulfuric acid is first heated by a short length
of copper tubing whose temperature is automatically
cycled. The temperature program in use at present
is to start the volatilization cycle with the tube
at about 70°C, raise the temperature to about 200°C
and hold it there for a length of time dependent on
the cycle time, and then shut off the heat a few
minutes before the end of the cycle so that the tube
temperature falls to 70°C by the time the next cycle
begins. This warm air flow serves to heat the
Teflon block pressing against the top surface of
the filter.

The lower part of the filter sealing mechanism
is Teflon-coated brass, which is heated to about
200°C. The volatilized acid is carried to the FPD
by a 1/8-inch TFE Teflon tube, which is maintained
at about 200°C. The methanol injector is a reservoir
for liquid methanol, which is maintained at a slightly

elevated, controlled temperature. The methanol
passes through a diffusion tube to the dry air flow.
A valve in the diffusion tube makes it possible to
shut off the methanol.

Filter Advance Mechanism

    The two-step monitoring method requires the
simultaneous use of two filters and a mechanism to
automatically cycle them between the ambient air flow
and the volatilization air flow. It is also necessary to have a mechanism that replaces the used
filters with clean ones after a preset number of
cycles to eliminate the possible interference due
to the nonvolatile particulates that collect on them.
The mechanism is shown in Figure 5.6.

Figure 5.6. Photograph of the filter changing mechanism.

The large gears and attached linkages open and
close the gas flow system so that the filters can
be moved. The linkage is spring-loaded to obtain a
relatively constant pressure seal at the filters and
to prevent damage if the mechanism attempts to close
on something thicker than a filter. The filters are
carried on a rotating arm driven by a 90° Geneva
mechanism.

When the pair of filters on the rotating arm
need to be replaced, the arm is rotated 90°, the
old filter is replaced by a clean one, and the arm
is rotated another 90° to carry the new filter into
the position for the ambient air flow. At the end
of the following cycle, the same action replaces the
other filter without any gap in the data.

The supply of clean filters is contained in a
tube below the rotating arm. A dual ratchet mechanism
lifts the stack of filters when a clean filter is to
be inserted, and this forces the new filter into
position in the rotating arm and the used filter
into a tube above the arm. The rotating arm and the
tube above it contain spring clips, so that a filter
can be pushed in from below, but will not fall out.

Electronics

The design philosophy for the prototype electronics
was to include as much flexibility as practical.
Some circuits, such as the heat programmer, were
designed so that parameters could be changed after
some operating experience had been obtained. The
electronic system is divided into four functional
sections: signal processor, timer, mechanical
function control, and heat programmer.

Signal Processor

The output of the FPD is a nonlinear function of
the sulfuric acid concentration in the air flow.
Therefore it is necessary to linearize the signal
before it is integrated to determine the total amount
of acid volatilized in a cycle. The integrator is
followed by a span-adjusting amplifier and a digital
panel meter. The amplifier can be set according to
the cycle time and the ambient air sampling rate so
that the digital readings have convenient units.
The digital panel meter provides a signal for a
digital printer, and holds the reading from a given

cycle until the next cycle is completed. Other output
signals available are: logarithm of the FPD output,
linearized FPD output, and integrator reading.

Timer

The timer provides the timing sequence for all
required functions. It operates from the line frequency so that the monitor will remain accurately
synchronized with the time of day. The time for a
complete cycle can be selected from 5, 10, 20, and
40 minutes or from any of these multiplied by factors of 3, 6 or 12. The multiplier does not increase
the length of time the FPD signal is integrated before being read. For example, when the time setting
is 20 minutes and the multiplier setting is 3, each
sample is collected over a period of one hour and
is analyzed for sulfuric acid during the first 20
minutes of the following cycle.

The timer can operate in either automatic or
manual mode. In the manual mode, the monitor automatically goes through one cycle each time it is
started. This mode of operation will be used when
calibrating the instrument, or when individual
filters with sulfuric acid already on them are being
analyzed. A cycle can also be manually interrupted
before it has run its full course.

Heat Programmer

This programmer controls the temperature of the
tube that heats the volatilization of air just before
the filter. The temperature of the tube is detected
by a thermistor. The circuit also reduces the heat
if the FPD signal becomes large enough to approach
the upper end of the range for which the calibration
is valid.

Mechanical Function Control

This circuit contains the microswitches that
sense the position of the mechanical components and
turn the motors on and off that drive the filter
advance mechanism. The circuit includes a counter
that can be set from 1 to 99, which determines the
number of cycles a pair of filters is used before
they are replaced by fresh ones. It also operates
the solenoid valve in the gas flow system.

Data

The prototype monitor has been run automatically
for extended periods of time to demonstrate that the
mechanical and electronic functions operate reliably.
The calibration of the apparatus is still in progress,
so it is not possible to report those results at this
time. The calibration methods are similar to those
used in the preliminary experiments. The response
of the FPD and the operation of the linearizer and
integrator circuits are calibrated by adding solutions containing known amounts of sulfuric acid to
the filter and controlling the conditions so that
the acid is volatilized at a relatively constant rate.
This produces records such as shown in Figure 5.4.
Both high and low rates of volatilization are used
to establish the slope of the lines like those in
Figure 5.1. The manufacturer's calibration of the
FPD gives a slope of 1.84 for SO_2, and the early
calibration data indicate that the slope for sulfuric
acid may be larger than that. The calibration of
the monitor will be checked by sampling aerosols
whose concentration has been determined separately
by electrolytic conductance.

We do not believe that it will be practical to
generate aerosols whose concentrations remain stable
to within the precision of the prototype monitor.
One way of assessing the precision of the instrument
is to sample repetitively from an isolated bag of
aerosol. In this case, the aerosol concentration
slowly decreases with time but does not vary
erratically. In the first test of this kind, which
used a 20 minute cycle time, the integrated output
reading was 175 in arbitrary units near the start of
the test. Successive readings were less by amounts
of 9, 11, 11, 11, 11, 12, 8, 8, 5, and 5, so that
the last reading in this series was 84. Thus the
output readings show little random scatter. This
series was terminated by moving the aerosol bag and
then obtaining successive readings of 85, 79, 72,
and 63. The discontinuity in the readings when the
aerosol was disturbed shows that the instrument was
responding to the aerosol concentration. The amount
of acid collected per cycle in this run was initially
somewhat less than 0.5 µg.

In this same test, a filter was removed from the
apparatus after a sample of aerosol had been collected
on it, but before the acid was volatilized. The
filter was set aside in a petri dish for 6 hours,
then placed in the monitor for volatilization. A
reading of 186 was obtained, whereas a reading

between 190 and 200 would have been expected if the
filter were analyzed immediately after collection.
It is believed that further tests of this kind will
show that aerosol samples can be collected on filters
with a portable sampler, and then brought to the
prototype monitor for automatic analysis.

The fact that the volatilization will not be
entirely complete in each cycle has already been
mentioned. The most important consequences of this
are a limitation on the smallest amount of acid the
monitor can reliably measure, and a tendency for
the result from each cycle to be influenced by the
amount of acid in the previous cycle. In the data
obtained so far, this latter effect has been small
enough that it should not present a problem. For
example, the reading of 186 mentioned earlier was
obtained when the previous reading was 63, and is
5 to 15 units lower than would have been obtained
if the previous reading were near 200.

The real test of the new monitoring method
described here will come in a few months when it is
used in the field. This work is being performed
pursuant to Contract No. 68-02-0592 with the
Environmental Protection Agency.

REFERENCES

1. Buechley, R. W., W. B. Riggan, V. Hasselblad, and J. B.
    VanBruggen. "SO$_2$ Levels and Perturbations in Mortality,"
    Arch. Environ. Health, 27, 134 (1973).
2. Barton, S. C. and H. G. McAdie. *Proc. of the 2nd Intern.
    Clean Air Congress* (New York: Academic Press, 1971), pp.
    379-382. Also, 164th National Meeting, Amer. Chem. Soc.,
    New York, August, 1972, Paper 73, Div. of Water, Air
    and Waste Chem.
3. Forrest, J. and L. Newman. "Ambient Air Monitoring for
    Sulfur Compounds," *J. Air Pollution Contr. Assn., 23,*
    761 (1973).
4. Richards, L. W. 165th National Meeting, Amer. Chem. Soc.,
    Dallas, April, 1973, Paper 47, Div. of Water, Air and
    Waste Chem.
5. Scaringelli, F. P. and K. A. Rehme. "Determination of
    Atmospheric Concentrations of Sulfuric Acid Aerosol by
    Spectrophotometry, Coulometry, and Flame Photometry,"
    Anal. Chem., 41, 707 (1969).
6. Dubois, L., C. H. Baker, T. Teichman, A. Zdrojewski, and
    J. L. Monkman. "The Determination of Sulfuric Acid in
    Air: A Specific Method," *Mikrochim. Acta,* 269 (1969).

7. Dubois, L., R. S. Thomas, T. Teichman, and J. L. Monkman. "A General Method of Analysis for High Volume Air Samples, I," *Mikrochim. Acta*, 1268 (1969).

8. LaMer, V. K., E. C. Y. Inn, and I. B. Wilson. "The Methods of Forming, Detecting, and Measuring the Size and Concentration of Liquid Aerosols in the Size Range of 0.01 to 0.25 Microns Diameter," *J. Colloid Sci.*, 5, 471 (1950).

CHAPTER 6

A COMPARATIVE REVIEW OF OPEN-PATH SPECTROSCOPIC ABSORPTION METHODS FOR AMBIENT AIR POLLUTANTS

W. A. McClenny, W. F. Herget, and R. K. Stevens

INTRODUCTION

Studies are now underway to improve the technical and scientific knowledge of atmospheric processes responsible for the formation, decay, and transport of pollutant gases on scales encompassing major urban centers, *i.e.*, 100-150 kilometers.[1] Included as part of these studies are plans to evaluate and improve current air quality simulation models that have been developed to predict the distribution of air pollutant concentrations in space and time.[2] Spatial averaging of these concentrations along 0.5 kilometer distances for central city predictions and increasing to 10 kilometers for rural areas is envisioned. To check these predictions and to provide input data to the simulation models, open-path monitors are expected to play a major role. These monitors typically employ a collimated beam of radiation and a remotely placed detector to measure changes in atmospheric transmission at wavelengths characteristic of the gas being measured. This paper is a review and comparison of various methods that have been proposed and/or used to provide measurement of path-averaged concentrations.

OPEN-PATH MONITORS

The ideal features for open-path monitors include those common to all ambient air monitors, such as specificity, sensitivity, and reliability. In addition, special features that apply to open-path monitors alone are: (1) the ability to make transmission measurements over long distances under various ambient conditions, (2) the ability to range, *i.e.*, to measure trace concentrations with a spatial resolution of some fraction of the total path length, and (3) the ability to make measurements in three dimensions. However, all these features may not be compatible with routine operation because either the open-path monitor must limit the energy density of the light beam to levels below those considered safe[3] or precautionary measures must be established that meet equivalent safety criteria.

Specificity requirements can usually be met by measurements at selected wavelengths in the ultraviolet, visible, or middle infrared portions of the electromagnetic spectrum. These selections are made to correspond to unique features (usually maxima and minima) in the absorption spectra of the gas being measured. Transmission at the selected wavelengths provides data from which trace concentrations can be calculated. For a nonturbulent atmosphere the sensitivity of a system for a given molecular species and path length is dependent on the noise levels inherent to the source, detector, associated electronics, and background radiation. Common solutions include the use of phase sensitive detection or signal-gated detection (pulsed sources) and the use of cooled detectors.

Measurements in the open air are subject to the effects of atmospheric turbulence. Local variations in temperature, due to warming of the earth and subsequent heat transfer to the atmosphere and to the effect of mixing caused by winds, result in a corresponding variation in the index of refraction along the measurement path. These local variations can be thought of as causing a focusing or defocusing effect over the light beam cross section.[4] If the local variation is smaller than the diameter of the light beam, the dominant effects are beam spreading and beam scintillation. Beam spreading is caused by small angle scattering; its effect is to increase the beam divergence and cause a decrease in the signal recorded by the detector. Beam scintillation is caused by small scale destructive and constructive

interference within the beam cross section, which
produces a corresponding variation in the spatial
and temporal power density over the radiation detector. If the local variation is larger than the
diameter of the light beam, beam deflection occurs.
Kerr[5] has shown that refractive index variations
decrease with increasing frequency and are negligible
at 1.0 kHz. This fact forms the basis for attempts
to eliminate the effects of turbulence by completing
a measurement in a time short compared to 1.0 msec,
thereby "freezing" the atmosphere.

Another consideration that applies to open-path
monitoring is the extent to which neutral attenuation
by haze, fog, dust, or other particulate matter and
attenuation by permanent atmospheric gases occur.
For example, absorption by H_2O and CO_2 limit the
useful regions of the middle infrared to several
"atmospheric windows," *i.e.*, spectral regions over
which the absorption by permanent atmospheric gases
is relatively weak.

Open-path monitors can be classified into two
basic groups: those using system elements at both
ends of the measurement path and those that are
single-ended. These two configurations are shown
schematically in Figure 6.1. A second classification

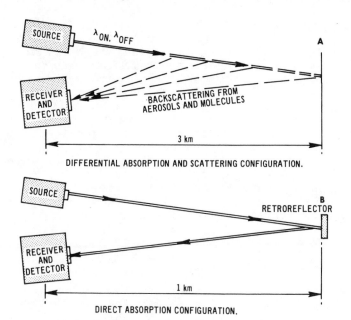

Figure 6.1. Typical open-path configurations.

into types based on radiation sources has two basic
subdivisions, broadband sources and laser sources.
The laser sources are tunable in wavelength either
continuously or in discrete steps. Five sources
have been chosen for comparison. As shown in Table
6.1, these are the $C^{12}O_2^{16}$ gas laser, the semiconductor
diode laser, the optical parametric oscillator (OPO),
the organic dye laser, and a representative broadband source.

Laser Sources, Doubled-Ended

For systems in which the wavelength interval of
source radiation is small compared to the linewidths
or absorption gradients in the absorption spectra of
atmospheric gases, the initial source intensity I_o
centered at wavelength λ is attenuated exponentially
in traversing a total distance L through the atmosphere. The signal intensity returned to the
transceiver in the absence of atmospheric turbulence
is:

$$I = KI_o \exp\left[-\alpha_o L - \sum_m \int_{\ell=0}^{L} n_m \sigma_m \, d\ell\right] \qquad (1)$$

where K denotes the transfer efficiency of the optics,
and α_o is the attenuation coefficient due to all
causes other than absorption by m molecular species
of density n_m and cross section σ_m. Measurements at
a total of m+1 different wavelengths give sufficient
information to obtain a unique answer for a particular
gas density averaged over a pathlength L, provided
α_o is independent of wavelength. Appropriate selection of the wavelengths to be used minimizes the
number of gases that are interferences as well as
the extent to which the interferences cause attenuation of the signal. If only two wavelengths are
used, the average value of the gas density is:

$$n = L^{-1} (\sigma_2 - \sigma_1)^{-1} \ln(I_1 I_{02}/I_{01} I_2) \qquad (2)$$

For more than two wavelengths, the concentration of
the target gas is a linear combination of terms of
the type $\ln(I/I_o)$, *i.e.*, a weighted average of the
logarithms of normalized signal returns.

Table 6.1

Characteristics of Sources Used in Open-Path Absorption Measurements

Source	CO$_2$ Gas Laser	Diode Laser	OPO	Dye Laser	Broadband
Spectral coverage	middle IR	middle IR	VIS–middle IR	UV – near IR	UV – far IR
Wavelength selection	intra-cavity dispersion	cavity length variation	phase matching	intracavity dispersion	dispersive and filter techniques
Average power output during operation (watts)	0.5 – 1.0	$10^{-3} - 10^{-2}$	$10^2 - 10^3$	$10^4 - 10^6$	1 cm^{-2}ster$^{-1}\mu^{-1}$
Typical pulse time (sec)	10^{-3}	10^{-6}	10^{-7}	$10^{-6} - 10^{-7}$	continuous
Output bandwidth [a],[b]	doppler width	10^{-4} cm^{-1} at constant T	0.25 cm^{-1}	1 Å	0.1–5.0 cm^{-1} at detector
DIAL capability	no	no	yes	yes	no
Resonance absorption capability	no	yes	yes	yes	yes
Coincidence absorption capability	yes	yes	yes	yes	yes
Special requirements	none	cryogenic cooling	laser coolant	none	none
Approach towards scintillation	freeze atmosphere	freeze atmosphere	freeze atmosphere	freeze atmosphere	spatial or time averaging
References	6, 8, 14	15, 16, 17	33	33, 34	19, 21

[a] Outputs from the dye laser and OPO can be narrowed by the addition of an intracavity etalon.
[b] For a diode laser pulse duration of approximately 1 μsec, the change in frequency (due to temperature) is 0.1 cm^{-1} so that the effective bandwidth must be taken as 0.1 cm^{-1}.

The use of the CO_2 laser for trace gas detection was suggested by Jacobs and Snowman[6] and by Hanst and Morreal.[7] The reduction to practice has resulted in a multiwavelength laser system for detection of ethylene, ammonia, ozone, and carbon dioxide as reported by Snowman and Gillmeister.[8] This system makes direct absorption measurements at each of four wavelengths in an open-path configuration similar to that shown in Figure 6.1b. The CO_2 wavelengths are spatially dispersed by an intracavity diffraction grating, and different wavelengths are obtained by use of a mechanical chopper that sequentially exposes mirrors placed beyond the grating, thereby completing the resonant cavity of the laser. A schematic of this system is shown in Figure 6.2. A system incorporating two frequency-stabilized CO_2 lasers operating simultaneously is being tested by Hildago and Christy. They have measured the absorption cross sections of O_3 at CO_2 laser wavelengths P(10) to P(36) in the 9 micron band. Assuming a 1% absorption differential and no interferences, the values for the pair of

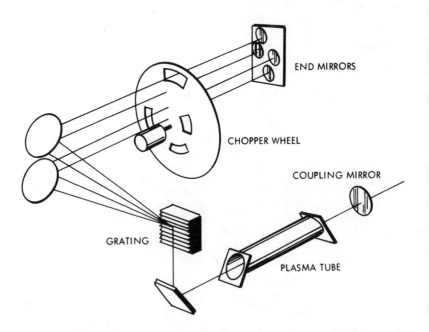

Figure 6.2. Wavelength selection process for CO_2 laser.

lines, P(14) (9.504µ) and P(24) (9.586µ), indicate a sensitivity of 0.01 ppm (20 µg/m^3) of O_3 over a total path length of 2.0 km.

Other gas lasers have coincidences with absorption bands of pollutant gases.[10] Measurements of absorption techniques can be applied in the case of NO, NO_2, and SO_2.[11] Comparisons between line positions and absorption spectra of various gases indicate that the HeXe laser can reach CO and H_2CO. Similar predictions of coincidences between CO and doubled CO_2 laser wavelengths and features in the absorption spectra of NO, NO_2, SO_2, and CO have been made.[12] Tuning of gas laser lines by Zeeman splitting has been used by Linford[13] to move Xe laser wavelengths several Ångstroms through formaldehyde absorption features.

For a prototype gas laser system, accurate values of the absorption coefficients and their dependence on ambient conditions of temperature and pressure (and for the case of water vapor, on relative humidity[14]) must be used to obtain accurate values for path-averaged densities. The prototype system would deal with atmospheric turbulence by completing a measurement in a time interval short compared to the temporal variation of the index of refraction or by signal averaging over appropriately long time intervals. To the extent possible, variation of the energy distribution of different wavelengths across a phase front at the transmitter[14] optics must be eliminated. Even if this is done, spatial intensity variations resulting from diffraction are likely to cause a variation from wavelength to wavelength in the signal intensity returned to a finite aperture receiver.

Another source for direct absorption experiments is the semiconductor diode laser. The spectral interval over which these tunable sources operate is determined by the choice of materials used in their fabrication. Figure 3.1 of Chapter 3 in this volume shows the range over which different materials operate. Choice of a specific value for the composition variable determines a tunable range of several wavenumbers. Within this range the diodes are skip-tunable from one continuously tunable longitudinal mode to another by varying the current applied to the diode.

The application of tunable diodes to air pollution problems has been discussed by Hinkley and Kelly[15] and Hinkley,[16] and direct absorption measurements of SO_2 across industrial stacks have been demonstrated.[17] In this latter application, a liquid helium-cooled

$Pb_{1-x}Sn_xTe$ diode operates in the 1120-1130 cm^{-1}
region corresponding to the ν_1 band of SO_2. Current
passed through the diode controls its performance.
A high current pulse causes laser emission at λ_1.
This pulse is followed by a steady low current that
resistively heats the diode to a higher temperature,
thereby changing the refractive index of the diode
and its optical length. A second high current pulse
causes laser emission at λ_2. The pulsed outputs
(typically 0.5 mW) at λ_1 and λ_2 are used in a direct
absorption system. Extension of the technique for
path lengths on the order of a kilometer appears
feasible.

Broadband Sources, Double-Ended

     In the interpretation of atmospheric transmission
studies using broadband sources, quantitative determination of trace gas concentrations is complicated
by the fact that an integrated cross section for
absorption must be considered, *i.e.*, the wavelength
interval of source radiation is large compared to
the linewidths in the absorption spectra of the gas
being measured. An example of this is the treatment
by Campani *et al.*[18] of effective integrated cross
sections in the infrared.
     Long path measurements of ozone spectra using
ultraviolet radiation from arc sources have been reviewed by Stair.[19] These early experiments measured
absorption at wavelengths centered at 2500 Å over
distances of approximately 0.3 km. More advanced
systems employing correlation techniques have developed in recent years. One of these, the Barringer[20]
COSPEC III,* measures SO_2 and NO_2. The instrument
consists of a telescope to collect radiation from a
remotely placed Xenon arc source, a two-grating
Ebert-Fastie monochromator for dispersion of the
incoming light, a disc-shaped multiple-slit mask and
an electronics system. As the mask rotates through
the exit plane of the monochromator, different slit
arrays allow selected groups of narrow wavelength
bands to be transmitted to the detector, thereby
providing sequential correlation in a positive and
negative sense with absorption bands of the target
gas. The detection of the corresponding difference
signal provides measurement of the path-averaged
concentration of the target gas.

---

*Mention of commercial products does not constitute an endorse-
ment by the U.S. Government.

Kay[21] has developed a correlation method that
involves matching the periodic passbands of bire-
fringent crystals with features in absorption spectra.
High energy throughput is obtained, although for NO_2
the absorption peaks are essentially periodic only
over restricted wavelength intervals. Calibration
of this instrument as well as the COSPEC III is
established by placing cells containing known amounts
of the target gas in the measurement path.

EPA is currently using a scanning spectrometer
with a continuum light source and telescopic optics,
as described in Chapter 13 of this volume, to gather
data on various pollution sources and to select
optimum wavelength regions for analysis of specific
pollutants. Major system components are a 1700° K
blackbody source, 60-cm diameter telescopic optics,
quarter-meter grating monochromator, and a closed-
cycle, cryogenically cooled Hg:Ge detector. In
operation over a four kilometer path the instrument
is capable of 1.0 cm^{-1} resolution in the 9 micron
spectral region and has been used to follow daily
ozone concentration variations of 0.05 to 0.2 ppm
(50 to 200 $\mu g/m^3$).[22]

Systems employing the technique of gas cell (or
gas filter) correlation spectroscopy are currently
being developed under funding by EPA and others.
In this technique high sensitivity and specificity
are achieved by modulating the light beam with cells
containing the gases to be detected. A detailed
description of the technique is given in Chapter 11
A simple form of the gas cell correlation system is
shown in Figure 6.3. The source is generally a
broadband source, but it could be a line source such
as a heated sample of the gas to be detected. The
correlation cell contains the gas to be detected in

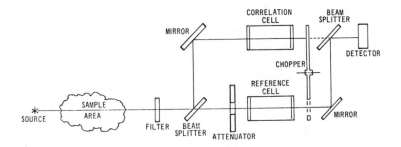

Figure 6.3. Simplified gas filter correlation system.

an amount sufficient to produce 100% absorption at
the line centers across a spectral range isolated
by the filter. The reference cell contains a nonabsorbing gas such as N_2. Either *in situ* or open-
path measurements can be made, *i.e.*, the sample area
can be a cell or an open path. The chopper allows
the detector to alternately see radiation transmitted
by the correlation cell and the reference cell. When
the sample area is free of the gas to be detected,
the instrument is balanced by adjusting the attenuator so that equal intensities reach the detector
through each cell. When the gas to be detected
enters the sample area, the energy transmitted by
the correlation cell will be essentially unchanged
since there is already 100% absorption at the line
centers; the energy reaching the detector through
the reference cell will drop due to the absorption
in the sample region. The resulting pulsed signal
is proportional to concentration for low concentrations. Particulates or other gases affect each beam
in the same way and produce no pulsed signal at the
detector. Two such systems have been fabricated for
EPA, one for auto exhaust measurements and one for
cross-stack applications. The extension of the
technique to open-path measurements is straight-
forward and will be carried out by EPA during FY 75.
Concentrations in the ppb range should be detectable
over paths of a kilometer or so.

Laser Source, Single-Ended

The remaining two categories of sources, the OPO
and dye laser, in addition to being tunable over the
ranges noted in Table 6.1, are pulsed sources with
a high energy per pulse. These features allow remote
sensing by a lidar-type technique represented schematically in Figure 6.1a. This technique, called
Differential Absorption LIDAR (DIAL), has developed
from the pioneering efforts of Schotland[23-25] and
has been recently predicted[26-29] to be capable of
measuring less than one ppm of NO_2, CO, and SO_2 with
a spatial resolution of 15 meters and a range exceeding one km. In operation, a wavelength tunable
laser pulse (10-100mJ) is backscattered from
atmospheric aerosols (Mie scattering). The backscattered radiation is collected with telescopic
optics and monitored with a sensitive detector.
The ratio, R, of this signal to the transmitted
pulse signal is such that,

$$R \propto \exp \left[-2 \left(\alpha_o L + \sum_m \int_{\ell=0}^{L} n_m \sigma_m \, d\ell \right) \right] \tag{3}$$

where L is the distance to the backscattering pulse, α_o is the scattering coefficient, and σ_m denotes the absorption coefficient of one of m molecular species. If no interferences are present ($m = 1$, and σ_o is independent of wavelength), only two wavelengths need be used to determine the target gas density. One, λ_1, would usually be chosen to coincide with an absorption maximum; the other, λ_2, with a minimum. The average density over distance L is then

$$n \equiv L^{-1} \int_{\ell=0}^{L} n(\ell) \, d\ell = L^{-1} (\sigma_2 - \sigma_1)^{-1} \ln (R_1/R_2) \tag{4}$$

where subscripts 1 and 2 refer to quantities associated with λ_1 and λ_2. By differentiating with respect to distance, ℓ, the spatial profile of concentration is obtained.

As seen from comparison with Equation (2), the average density as specified in Equation (4) is in the same form as that for direct absorption. However, R now denotes the ratio of transmitted to received beam energies. Comparison of DIAL and direct absorption shows that DIAL provides ranging capability by time of flight measurement, spatial resolution, and a three-dimensional, single-ended measurement capability. Direct absorption is simpler and can use many of the low power laser and broadband sources presently available.

Use of the DIAL technique to measure pollutant gases was first reported by Walther in Germany.[30] An organic dye laser was used to detect NO_2 over paths up to four kilometers. Concentrations in the 0.25 ppm range were measured. The technique is being pursued most vigorously in this country at Stanford University and Stanford Research Institute under a grant from the National Science Foundation.

Zaromb[31] has proposed a technique similar to DIAL in which the backscattered signal is due to Raman scattering from O_2 and N_2. In this case the laser pulse generates a return at a number of wavelengths that are dispersed at the receiver prior to detection. If two of these wavelengths lie along a gradient in the absorption spectrum of the target gas, relative signal returns can be processed to give information equivalent to that of DIAL. Byer and Garbuny[28] have analyzed in considerable detail the use of topological targets as retro-reflectors in direct absorption systems. They predict ranges of 3-10 kilometers for a 10mJ transmitted pulse energy with a detection sensitivity below 0.1 ppm. Garbuny and Henikson[32] have already established experimentally that backscattering from a topological target at 100 meters can be used to detect first overtone absorption by a confined sample of CO.

SUMMARY

Of the systems compared in Table 6.1 for double-ended measurements, those using broad band sources, the CO_2 laser, and the tunable diode laser are the most advanced. Although the COSPEC instruments have been used mainly with scattered sunlight as the source, recent long line experiments[33] have indicated the feasibility of obtaining ambient air measurements of as low as 20 ppb of NO_2 over total distances of 0.67 km. However, calibration and zero drift problems have not been resolved. The tunable diodes, while offering the greatest wavelength range of the laser sources, require cryogenic cooling for both diode and detector. The CO_2 laser system does not require special coolants, but other gas lasers (HeNe, HeXe) must be used to provide wavelength coverage matching the projected versatility of the diode. Of all the double-ended techniques, that employing gas-cell correlation may well prove to combine the greatest sensitivity, specificity, and simplicity with least cost. However, the technique is restricted to gases possessing line spectra and remaining stable in a closed cell.

For the DIAL technique, the OPO tunable lasers have the greatest potential for the infrared spectral region, while dye lasers are practical for use now in the visible and ultraviolet.[35] At present the parametric generation of tunable radiation is limited to the use of $LiNbO_3$ as the host crystal. This limitation restricts the detection scheme to the near infrared overtone and combination band absorption of various pollutants with attendant loss of sensitivity. The use of organic dyes such as 7-diethylamino-4-methylcoumarin has the disadvantage of thermal decomposition of the dye upon repeated exposure to flashlamp pulses.

REFERENCES

1. Proceedings of the Third Meeting of the Expert Panel on Air Pollution Modeling, Paris, France, October 2-3, 1972, p. VII-1.
2. Proceedings of the Third Meeting of the Expert Panel on Air Pollution Modeling, Paris, France, October 2-3, 1972, p. VI-1.
3. Clark, A. M. "Ocular Hazards," in *Handbook of Lasers with Selected Data on Optical Technology.* (Cleveland, Ohio: The Chemical Rubber Co., 1971).
4. Davis, J. I. *Appl. Optics, 5,* 139 (1966).
5. Kerr, J. R. *J. Opt. Soc. Am., 62,* 1040 (1972).
6. Jacobs, G. B. and L. R. Snowman. *IEEE J. of Q. E.,* QE-3, No. 11, Nov. 1967, pp. 603-605.
7. Hanst, P. L. and J. A. Morreal. *J. Air Poll. Control Assoc., 18,* 754 (1968).
8. Snowman, L. R. and R. J. Gillmeister. in *Joint Conference on Sensing of Environmental Pollutants, Palo Alto, Calif.* (New York: American Institute of Aeronautics and Astronautics, November 1971), Paper No. 71-1059.
9. Hidalgo, J. U. and E. H. Christy. Interim Report, EPA Grant No. 801429 (Environmental Protection Agency, Research Triangle Park, North Carolina, 1973).
10. Hanst, P. L. in *Advances in Environmental Science and Technology,* Vol. II, J. N. Pitts and R. L. Metcalf, Eds. (New York: John Wiley & Sons, 1971), pp. 91-213.
11. Gillmeister, R. J. G. E. Company, Pittsfield, Mass., private communication.
12. Menzies, R. T. *Appl. Opts., 10,* 1532 (1971).
13. Linford, G. J. *Appl. Opts., 12,* 1130 (1973).
14. Final Report: "Field Study on Application of Laser Coincidence Absorption Measruement Techniques," Contract EHSD 71-8, February 1972, prepared for the Environmental Protection Agency by G. E. Electronics Laboratory, Syracuse, New York.
15. Hinkley, E. D. and P. L. Kelly. *Science, 171,* 635 (1971).
16. Hinkley, E. D. *J. Opto-Electronics, 4,* 69 (1972).
17. Hinkley, E. D. Final Report, EPA Contract No. 68-02-0569 (Environmental Protection Agency, Research Triangle Park, N.C., 1973).
18. Campani, P., C. S. Fang, and H. W. Prengle, Jr. *Appl. Spect., 26,* 372 (1972).
19. Stair, R. in *Ozone Chemistry and Technology,* Advanced Chemistry Series, Vol. 21 (Washington, D.C.: American Chemical Society, 1959), pp. 269-285.
20. Barringer, A. R. and J. H. Davies. Presented at Joint Conference on Sensing of Environmental Pollutants, Palo Alto, California, November 1971, Paper No. 71-1105. (Available from Barringer Research, Rexdale, Ontario, Canada.)

21. Kay, R. B. Interim Report, EPA Research Grant No. 800708 (Environmental Protection Agency, Research Triangle Park, N.C., 1973).

22. Streiff, M. L. and C. B. Ludwigg. Final Report, EPA Contract No. 68-02-0020 (Environmental Protection Agency, Research Triangle Park, N.C., 1973) (Report No. EPA - 650/2-73-026).

23. Schotland, R. M. *Proc. Third Symp. on Remote Sensing of the Environment* (Ann Arbor, Michigan: University of Michigan, October, 1964), pp. 215-224.

24. Schotland, R. M., A. M. Nathan, E. E. Chermack, and E. E. Uthe. Technical Report 2, Contract DA-36-039SC-87299 Dept. of Meteorology and Oceanography, New York University, New York (1962).

25. Schotland, R. M., E. E. Chermack, and D. T. Chang. *Proc. First Symposium of Humidity and Moisture* (New York: Reinhold Book Division, 1964), pp. 569-582.

26. Measure, R. M. *Opto-Electronics, 4,* 141 (1972).

27. Wright, M. L., E. K. Proctor, and E. M. Measure. Fifth Conference on Laser Radar Studies of the Atmosphere, Williamsburg, Va. (Sponsored by Amer. Met. Soc. and Opt. Soc. Amer., June 1973).

28. Byer, R. L. and M. Garbung. *Appl. Opt., 12,* 1496 (1973).

29. Ahmed, S. A. *Appl. Optics, 12,* 901 (1973).

30. Rothe, K. W., V. Brinkmann, and H. Walther. Proceedings of the Eighth International Conference on the Physics of Electronic and Atomic Collisions, Belgrade, July 16-30, 1973.

31. Zaromb, S. *1969 Proceedings of the Electro-Optical Systems Design Conference* (Chicago: Industrial and Scientific Conf. Management Inc., 1970), pp. 699-624.

32. Private communication from Dr. Max Garbuny of Westinghouse Research Laboratories, in Pittsburgh, Pa.

33. Contract Report and Supplement for EPA Purchase Order No. 3-02-04987.

34. Chromatix, 1145 Tena Bella Avenue, Mountain View, California 94040.

35. Dewry, C. F., Jr. *Modern Optical Methods in Gas Dynamic Research,* Dosanjh, Ed. (New York: Plenum Press, 1971), pp. 221-270.

SECTION II

TECHNIQUES TO MEASURE CHEMICAL AND PHYSICAL
PROPERTIES OF PARTICLES IN THE ATMOSPHERE

CHAPTER 7

APPLICATION OF X-RAY FLUORESCENCE TECHNIQUES
TO MEASURE ELEMENTAL COMPOSITION OF PARTICLES
IN THE ATMOSPHERE*

Joseph M. Jaklevic, Fred S. Goulding,
Blair V. Jarrett and John D. Meng

The design and operation of a system for the
automated X-ray fluorescence analysis of atmospheric
particulates on filters will be described. Central
to the system is a low-background Si(Li) semiconductor
detector coupled to a low-power variable-energy X-ray
tube used to generate monoenergetic photons for
fluorescence excitation. Sequencing of the filter
analysis, changes in X-ray excitation energy and
intensity, together with data storage can be performed without operator intervention under the
control of a small computer. Reduction of pulse
height data and intensity calibration can be performed in real-time in this same computer. The three
analyses required to cover over 30 elements can be
performed in a total time of 30 minutes with detection limits in the worst case of 25 ng/cm^2 or less
over filter areas of 7 cm^2. A single analysis for
a restricted group of 15 elements at a detection
level below 10 ng/cm^2 can be performed in ten minutes.
Samples are generated using a fully automated remote
sampling station designed for the unattended collecting
of up to 30 filter samples using a variety of commercially available filtering materials. The accuracy
and sensitivity of the method compared to other
sampling and analysis techniques will be discussed.

*Work supported in part by the Environmental Protection Agency
under Interagency Agreement with the U.S. Atomic Energy Com-
mission Contract No. EPA-IAG-0089(D)/A.

INTRODUCTION

X-ray fluorescence is an analytical technique in many ways ideally suited for the routine elemental analysis of atmospheric particulates. The good sensitivity, accuracy and speed, combined with low cost per analysis make it more than competitive with other available methods in applications involving large numbers of samples. A typical sample consists of a uniform deposit of small particles (<100-μm diameter) collected on a clean filter backing. This is almost an ideal thin specimen and permits accurate calibration of X-ray fluorescence measurements. Analysis is nondestructive and, in the case of energy-dispersive X-ray analysis, many environmentally important elements can be measured simultaneously using a single calibration.

An automated air-particulate analysis system, based on energy-dispersive X-ray spectroscopy,[1] is capable of the unattended analysis of over 30 filter samples for up to 35 elements on each sample. A complete analysis cycle for each filter requires 30 minutes, although analysis for 15 environmentally significant elements can be performed in only 10 minutes. No sample preparation is required, only long term calibration checks are necessary, and the analysis results are available immediately.

A brief discussion of the basic energy dispersive spectroscopy method and its applicability to the instrument design follows. For more detailed accounts of X-ray fluorescence techniques see References 2 and 3; for accounts of the energy dispersive method, see References 4 and 5.

PHOTO-EXCITED ENERGY-DISPERSIVE
X-RAY FLUORESCENCE

The X-ray fluorescence method is illustrated in Figure 7.1. Incoming radiation interacts in the sample to produce vacancies within the inner shells of the atoms of interest; these vacancies then de-excite with the accompanying emission of fluorescent X-rays whose energies are characteristic of elements in the sample. The term "photon-excited" refers to the character of the exciting radiation which, in the present case, consists of X-ray photons of sufficient energy to ionize the atomic shells in elements of interest. Alternative methods of

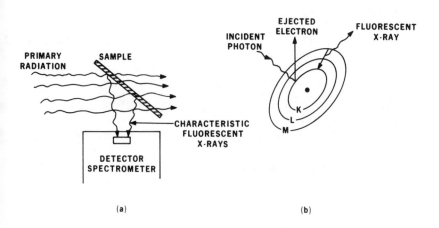

*Figure 7.1.   Illustration of the principles of an X-ray
fluorescence analyzer.*

excitation include electrons and positive ions.
Although the latter has been proposed for a number
of environmental monitoring purposes, in general its
capabilities are inferior to photon-excitation with
the exception of some rather specific sample forms.[6,7]
   "Energy-dispersive" refers to the method of
measuring the X-ray energies, distinguishing it from
the more traditional wavelength-dispersive techniques.[2,3]   In the energy-dispersive method, a semiconductor radiation detector measures the free charge
produced in the ionization cascade following the
photoelectric interaction of the X-ray within the
active volume of the semiconductor crystal.   This
charge is proportional to the energy of the original
X-ray.   Resulting signals can be used as a measure
of the energies of the incident X-rays over a wide
energy range.   This multiple-element analysis
capability constitutes an important advantage of
the energy-dispersive method in elemental contamination surveys and in studies of the interaction of
combinations of elements in pollution studies.
   Figure 7.2 is a schematic of the X-ray spectrometer
system, including detector and processing electronics.
The lithium-drifted silicon detector, normally 3-5 mm
thick and 0.5-1 cm in diameter, is cooled to liquid-
nitrogen temperature, primarily to reduce its leakage
current (resulting from thermally generated charge
carriers) and the associated electrical noise to a
very low value.   Charge carriers (holes and electrons)

result from the absorption of X-rays on the detector and are collected by applying an electric field across the active volume of the detector. The signal current through the detector is integrated by the first stage of the amplifier and the resulting step-function signals are amplified and shaped in a main amplifier unit before being fed to a pulse-height analyzer (or computer). The pulse height spectrum then represents a histogram of the number of events of various X-ray energies incident on the detector.

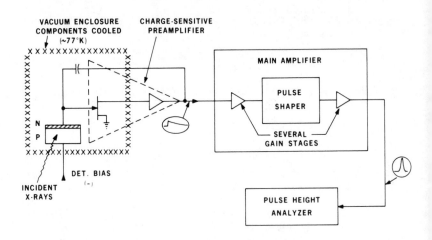

Figure 7.2.   The overall detector-electronics system used in an X-ray fluorescence analyzer.

The energy resolution of the system determines its capability to resolve X-rays from adjacent elements in the periodic table, and also affects the detectable limit for analysis of elements in the presence of background. The resolution is dependent both in the basic ionization process in the detector and on the signal/noise capabilities of the pulse processing electronics. Advances in the latter areas have made energy-dispersive X-ray analysis possible. Modern spectrometers are capable of energy resolution more than adequate for resolving the characteristic *K* X-ray of those elements having atomic numbers greater than sodium. Analytical sensitivities in the <1 ppm range are routinely achieved for many elements, although the general question of detectability is complicated by the possible presence of overlapping lines from more than one element.

The pulse-height spectrum observed in fluorescence analysis contains peaks of the characteristic X-ray of interest together with a large number of events due to the scattering of the primary beam in the region of the sample. This scattering can occur either without energy loss by the primary photon (Rayleigh) or with a small energy loss, depending on the scattering angle (Compton). Figure 7.3 is an idealized spectrum observed by irradiating a sample with monoenergetic primary photons. The two large peaks correspond to the scattering discussed above, the distribution at low energies is caused by scattering occurring in the detector itself. Fluorescent X-rays are evident in the middle of the spectrum, superimposed on a low background due to numerous processes that can contribute to partial amplitude signals (*e.g.*, multiple scattering, incomplete charge collection in the detector). A substantial reduction in background can be realized using a guard-ring detector with special pulse-rejection circuitry.[8] This unique system eliminates many of the causes of incomplete charge collection that can occur in the detector.

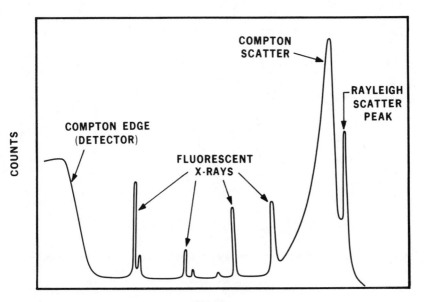

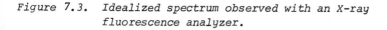

Figure 7.3.   *Idealized spectrum observed with an X-ray fluorescence analyzer.*

The use of monoenergetic excitation results in
the lowest background level in the region of interest.
This form of primary radiation can be approximated
by the characteristic K_α and K_β X-ray spectrum of a
selected target. In the present system, a target is
irradiated with the primary radiation from a tungsten
anode X-ray tube and the fluorescent X-rays from the
target excite the sample. This "secondary fluorescence" geometry allows the exciting X-rays striking
the sample to be varied in energy by changing the
secondary fluorescence target. This technique can
be used to compensate for the dependence of excitation efficiency upon the difference between the energy
of the characteristic X-ray and the primary radiation.
Figure 7.4 shows plots of the relative probability

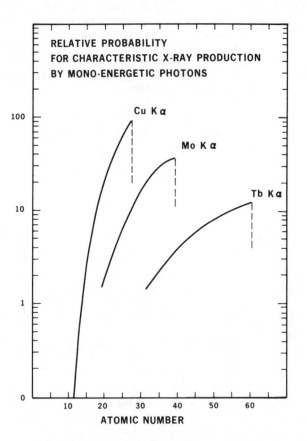

Figure 7.4.   *Calculated relative K X-ray production yields for
three excitation energies (Cu Kα, Mo Kα, Tb Kα
X-rays).*

of excitation of different elements for the three
secondary targets selected for the present system.
The excitation probability is calculated from fundamental properties of ionization cross sections
and fluorescence yield and can be converted directly
to analytical sensitivity in the case of energy-
dispersive analysis. An important feature is the
smooth variation of the X-ray production probability
as a function of the Z of elements for each of the
curves. This implies that a calibration curve can
easily be interpolated between adjacent Z elements
in cases where calibration standards cannot be con-
veniently prepared. Furthermore, the shape of these
curves is not dependent on the X-ray tube-sample-
detector geometry; a measurement on a single element
standard corrects for the geometry factors and provides a sensitivity calibration for the whole range
of elements covered by one fluorescer target.

INSTRUMENT DESIGN

Figure 7.5 is a cross section of the spectrometer
illustrating the close-coupled X-ray tube/secondary
target/sample geometry. With these large solid angles
we are able to operate at maximum counting rates of
10,000 counts/sec using less than 40 watts of anode
power in the X-ray tube. The secondary targets can
be automatically switched to provide a variable
energy of fluorescence excitation; the present targets are Cu, Mo, and Tb. Over 30 air filters mounted
in special holders can be sequenced through a complete
analysis cycle without operator intervention.
Control and monitoring functions together with
data acquisition and analysis are carried out by a
small on-line computer. The secondary target sequence
and the dead-time corrected counting interval are
chosen via front panel switches. Once the "start"
button is pressed, the system will automatically
perform the desired analysis and print out the results in ng/cm^2 for each of the filters in the stack
loader. In addition to the printout, the system
also writes the original spectral data and the results
of the data reduction on magnetic tape.
Figures 7.6 and 7.7 are logarithmic plots of
X-ray fluorescence spectra acquired on the same air
filter in five-minute counting intervals using the
Mo and Cu fluorescence, respectively. Typical
concentrations for some of the observed elements

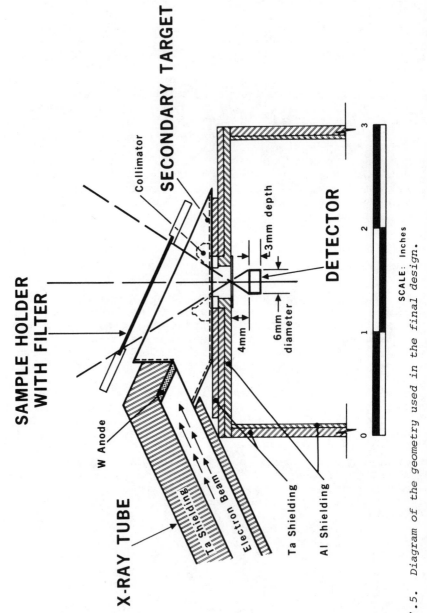

Figure 7.5.   Diagram of the geometry used in the final design.

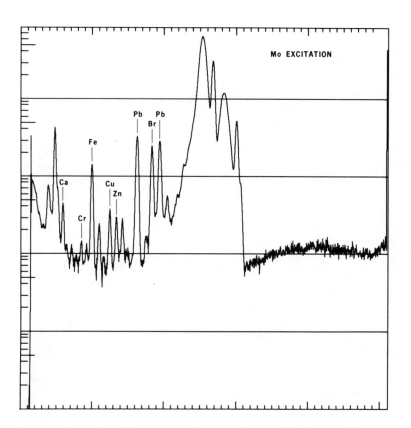

Figure 7.6.   Air filter   spectrum taken using the molybdenum
fluorescer.

are: PB-1.32 µg/cm², Fe-0.5 µg/cm², Mn-30 ng/cm²
and Ca-0.5 µg/cm².   The spectral response is similar
to that shown in Figure 7.3 with appropriate adjust-
ment for the presence of both a K_α and K_β peak in
the spectrum of exciting radiation.   The improved
peak-to-background for light elements is apparent
in the spectrum acquired using the Cu excitation.

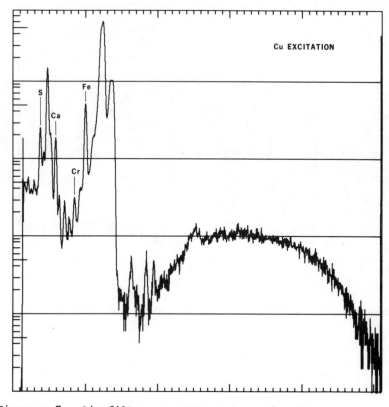

Figure 7.7. *Air filter   spectrum taken using the copper fluorescer.*

DATA ANALYSIS

The multiple-element capability of the energy dispersive method is illustrated in the large number of peaks in the spectra. For the method to be quantitative, the areas of individual peaks must be extracted from the data and an appropriate calibration applied. The principal problems associated with performing this computer analysis are subtraction of the continuous background from the region of interest and correction for interference between peaks of different elements due to the multiple structure of X-ray spectra. Both problems are solved by using a large memory in the computer to store spectra derived from a blank filter and from elements of interest. Experience shows that the background is relatively insensitive to the impurity concentration on a filter, so a blank filter can be run to provide a background spectrum that includes any impurity element either in the filter material or arising from the system itself.

Figure 7.8 shows a schematic sequence of the operations in the program. The original spectrum is sequentially reduced by first subtracting an appropriate multiple of the stored background compared over a selected region, and then by sequentially stripping out each of the spectra due to the individual elements. The factor determining the amount of each elemental spectrum to be subtracted, which is determined by the computer, is simply related to the area of the peak. Peak areas are then converted to concentrations in ng/cm by applying the appropriate excitation and detection efficiency function similar to those shown in Figure 7.4.

CALIBRATION

The expression given in Figure 7.9 is the formal mathematical expression of the intensity of the K X-ray for element i for the case of monoenergetic photons of energy E_O and intensity I_O incident on a slab of thickness d. The quantity p_id is the mass in ng/cm^2; other parameters are G, the solid angle, τ_i the photoelectric cross section for element i for photons of energy E_O, w_k^i the fluorescence yield for the K or L shell vacancies, ε_i the efficiency for detecting the fluorescent X-rays, and μ_O and μ_i the total mass absorption coefficients for the incident and fluorescent radiations, respectively. The quantity in brackets is, in effect, a correction for the absorption experienced by the radiation for thick samples. In most air filter applications it can be neglected except for very light element analyses $(Z = 20)$.

The equation can, in principle, relate the area of the peak (I_i) to the concentration (p_id) if the physical parameters are either measured or taken from literature values. In practice it proves more reliable to measure the complete function for a few selected elements and interpolate a smooth curve

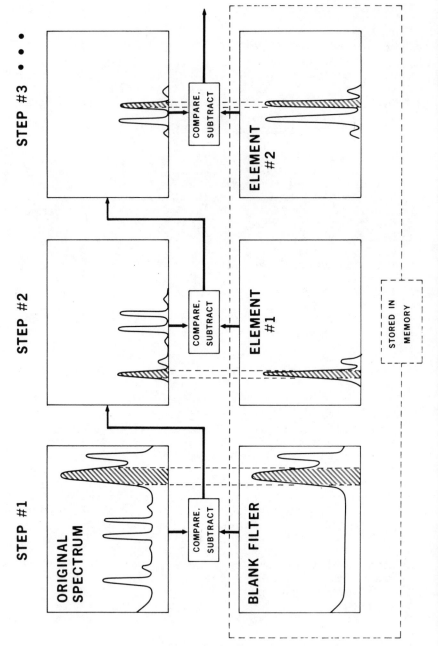

Figure 7.9. Illustration of spectrum stripping procedure.

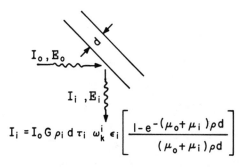

$$I_i = I_0\, G\, \rho_i\, d\, \tau_i\; \omega_k^i\; \epsilon_i \left[\frac{1 - e^{-(\mu_0 + \mu_i)\rho d}}{(\mu_0 + \mu_i)\rho d} \right]$$

Figure 7.9.   Expression for the overall efficiency of the process of production, absorption and detection of the fluorescent X-rays.

between the data for the remaining elements. Figure 7.10 is a plot of a family of curves obtained using a series of thin-film evaporated standards. The results can be compared directly with the purely theoretical results shown in Figure 7.4. The discrepancy for heavier elements is due to the rapid decrease in detector efficiency at higher X-ray energies, which was not included in the calculations. Since the shape of these curves is invariant for a fixed fluorescence-sample-detector geometry, subsequent calibrations only involve renormalization for variations in intensity of the exciting radiation, and can be achieved by measuring a single-element standard. Experience has shown that the stability of the system is sufficient to obviate the need for frequent calibrations.

Using the above curves, the validation results given in Table 7.1 were achieved. The majority of standard samples were in the form of thin evaporated films, although a number of elements were checked using compounds with known elemental ratios. These measurements were independent of any prior calibration data. The results include the automatic computer analysis. The agreement represents an average deviation of less than 5% over all elements.

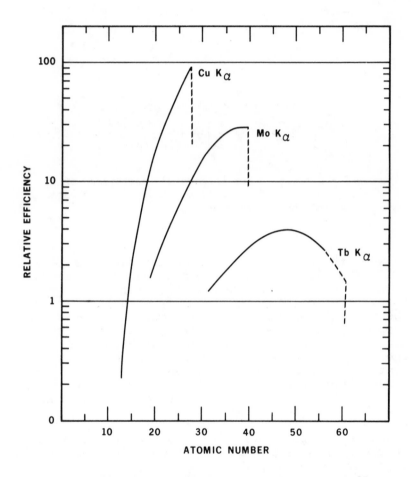

*Figure 7.10. Actual relative efficiency curves for three
fluorescers. These can be compared with the
theoretical curves of Figure 7.4.*

RESULTS

The performance of any analytical system is judged
from the demonstrated accuracy it can achieve with
realistic samples—in our case air filters. Table
7.2 is a comparison of the results obtained in the
automatic analysis of a series of eight air filter
samples taken on Millipore backings. The results

Table 7.1

Comparison of Measured Concentration of Standard Samples

Element	Fluorescer	Measured Density[a] ($\mu g/cm^2$)	Accepted Value[b] ($\mu g/cm^2$)
Al	Cu	1550[c]	2200[d]
Si	Cu	2380	2430[d]
S	Cu	2820	2970[d]
Ti	Cu	90	101
Cr	Cu	117	122
Fe	Cu	83[e]	83
Ti	Mo	100	101
Cr	Mo	121	120
Fe	Mo	94	87
Ni	Mo	109	100
Cu	Mo	55	49
Pb	Mo	132	131
Zr	Tb	66	61[f]
Pd	Tb	138	142
Cd	Tb	92	93[f]
Sn	Tb	142	138
Ba	Tb	122	124[f]

[a] Statistical errors are less than 1% in all cases of evaporated films.

[b] Thicknesses of evaporated films were determined by weighing. Estimated errors are <5%.

[c] The discrepancy in this comparison could easily be due to heavy elements in the 1100 Al alloy used.

[d] These densities represent the effective weight of infinitely thick samples.

[e] The Cu fluorescer comparisons are normalized to the Fe value.

[f] These were obtained by using samples of $ZrBr_4$, $CdBr_2$, and $BaBr_2$; the Br intensity was measured with the Mo fluorescer.

Table 7.2

Comparison of Results Obtained on Several Filter Samples

		Sample Number							
		1	2	3	4	5	6	7	8
Ca	(i)	261 ± 17	328 ± 18	333 ± 18	343 ± 18	298 ± 18	236 ± 17	199 ± 16	435 ± 20
	(ii)	271 ± 25	393 ± 25	361 ± 25	383 ± 44	286 ± 45	200 ± 43	177 ± 42	523 ± 52
Ti	(i)	30 ± 7	43 ± 12	35 ± 10	38 ± 12	27 ± 9	16 ± 7	19 ± 6	41 ± 12
	(ii)	37 ± 18	47 ± 20	48 ± 10	49 ± 20	42 ± 15	62 ± 20	43 ± 15	66 ± 15
Mn	(i)	< 5	7 ± 5	2 ± 5	26 ± 5	8 ± 5	< 5	< 5	22 ± 5
	(ii)	11 ± 4	6 ± 4	15 ± 4	27 ± 8	< 23	<22	<22	27 ± 8
Fe	(i)	410 ± 8	463 ± 8	421 ± 8	575 ± 8	319 ± 7	212 ± 6	180 ± 6	501 ± 8
	(ii)	405 ± 20	466 ± 23	421 ± 21	572 ± 23	371 ± 15	243 ± 10	180 ± 9	480 ± 19
Ni	(i)	8 ± 3	5 ± 3	< 3	< 3	< 3	< 3	< 3	8 ± 3
	(ii)	4 ± 2	8 ± 2	< 6	<11	<11	<11	<11	9 ± 4
Cu	(i)	97 ± 4	74 ± 4	83 ± 4	26 ± 3	4 ± 3	< 3	< 3	12 ± 3
	(ii)	81 ± 4	53 ± 3	55 ± 3	27 ± 4	7 ± 4	6 ± 4	6 ± 4	18 ± 4
Zn	(i)	59 ± 3	52 ± 3	43 ± 3	111 ± 3	23 ± 3	13 ± 3	10 ± 3	46 ± 3
	(ii)	58 ± 3	51 ± 3	41 ± 2	91 ± 4	24 ± 3	13 ± 3	12 ± 3	40 ± 3
Br	(i)	81 ± 3	157 ± 3	111 ± 3	170 ± 3	174 ± 3	129 ± 3	47 ± 3	394 ± 4
	(ii)	95 ± 5	175 ± 9	127 ± 5	171 ± 7	179 ± 7	129 ± 5	40 ± 4	408 ± 16
Pb	(i)	354 ± 8	680 ± 8	451 ± 8	763 ± 9	665 ± 8	511 ± 8	198 ± 6	1329 ± 11
	(ii)	327 ± 16	716 ± 36	449 ± 22	747 ± 30	702 ± 28	563 ± 23	196 ± 8	1320 ± 53

Errors reflect counting statistics only. In case where the quoted error is less than 10%, a 10% calibration error should be assumed.

(i) Values measured on EPA system   (ii) Values measured by R. Giauque.

marked (i) are the present measurements, whereas (ii) represent values obtained by carefully executed manual X-ray fluorescence measurements by R. Giauque in our laboratory. His system and method have been extensively checked and validated by alternative analytical methods.[4] The agreement between the two sets of measurements is within the quoted errors in most cases, a significant achievement when one considers that the present analyses were performed in less than one hour without operator intervention.

A series of separate validation results has been obtained for a number of elements using neutron activation as a comparison. General agreement between the two methods has been within 10%.[9]

The sensitivity of the technique for detection of elements present in very small quantities has been evaluated by measuring the minimum detectable limit for the various elements. This is defined as the quantity of material required to give a peak area equal to *three* times the statistical error in the background under the peak during a *five-minute* counting interval. These measurements also include the filter flow rate of our associated air samples, which was measured to be 0.8 m^3 of air/cm^2 of filter material over the two-hour sampling period. The minimum detectable levels shown in Figure 7.11 are quoted in ng/m^3 of air for the above samples. To convert to ng/cm^2 on the Millipore filter, the results should be multiplied by 0.8. To a first approximation these detection curves should correlate with the relative efficiency factors shown in Figure 7.10. Slight differences in the shape of the curves result from variations in the shape of the background for a given fluorescer; relative differences from one fluorescer to the next are also affected by the difference in the incident X-ray yield for each X-ray tube setting. In particular, this accounts largely for the values for detectability measured for Tb excitation being worse than would be expected on the basis of the calculated sensitivities.

An important consideration in comparing these detectable limits with results obtained for competing methods is the multiple-element detection capability of energy-dispersive X-ray fluorescence. The three curve segments in Figure 7.11 represent the sensitivities for *simultaneous* detection of many elements excited with each of the three fluorescers. (This statement is not rigorously accurate since it neglects reduction in detectability due to interelement

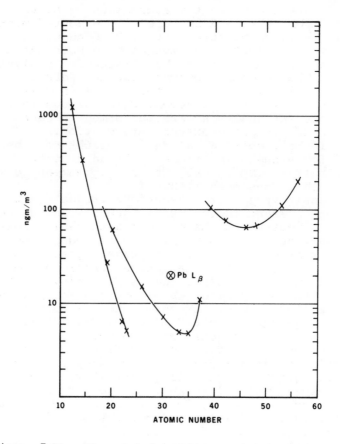

Figure 7.11. *Elemental detection sensitivity curves for the three fluorescers (two-hour sample collection time, five-minute analysis time on each fluorescer)*

interferences; however, in cases where the variation in concentration between adjacent elements is not large, the data are accurate enough for comparison.) It can be argued that by filtering the X-rays, the sensitivity could be optimized for one element. However, one of the greatest strengths of the energy-dispersive method is its multielement capability, which would appear to be important in environmental research and monitoring. Any comparison of the cost of analysis with that of competing analytical methods should bear this in mind.

PARTICLE SIZE AND MATRIX EFFECTS

As noted earlier, calibration problems arise for light elements due to the absorption of the low-energy characteristic X-rays either by the filter matrix or by the individual aerosol particles. Since the mean absorption length for these X-rays may be short compared to particle diameters, or to the filter thickness, the X-ray intensity reaching the detector depends upon the microscopic location from which the X-ray is generated either within the particle or in the filter matrix. Calculations of elemental concentrations using the observed X-ray intensity must then include a correction for this effect.

To calculate a reasonable correction factor, it is necessary to know something about the particle size distribution and the location of the intercepted particles within the filter material. Information concerning particle size must be obtained either by restricting the size range reaching the filter (*e.g.*, using impactors) or by making some assumptions regarding the size distribution in the original aerosols. Similarly, the absorption correction due to the filter matrix must be estimated by assuming localization of the particles in the filter, most likely on its surface.

At its best, any assumption appears to have dubious merit, so we have limited our investigation of the problem to estimating the maximum effect anticipated in certain limiting cases. This has been done by calculating the difference between the observed X-ray intensity with and without the absorption effects. Referring to Figure 7.9 we see that the difference between a thin film X-ray intensity and that including the absorption of the X-rays integrated over a thickness d is given by a factor:

$$A = \frac{1 - e^{-(\mu_0 + \mu_1)\, \rho d}}{(\mu_0 + \mu_1)\, \rho d} \tag{1}$$

where μ_0 and μ_1 are the total absorption coefficient for the exciting and emitted radiation, respectively. If we now associate d with the diameter of a homogeneous particle, we can calculate the absorption correction A as a function of particle size. (This calculation will overestimate the correction for spherical particles since it assumes a constant

thickness; however, since so little is known about
particle shapes, the assumption is as valid as any
other.) Figures 7.12, 7.13 and 7.14 are the results
of calculations for the case of Al, S and Ca X-rays
excited by Cu Kα radiation. The individual curves
represent various assumptions regarding particle
composition; the hydrocarbon assumes a unit density
material having the absorption cross sections of
carbon. The results indicate that no serious problems occur for particles of size below 10μ except
in the case of Al. Estimates of matrix effects can
also be obtained from these curves by recognizing
that the 5 ng/cm^2 Millipore filter is equivalent to
a 50μ-thick hydrocarbon sample. Thus, if the
material were uniformly distributed throughout the
filter, the correction to the intensity at its
maximum would be the value of the hydrocarbon absorption correction at 50μ. Again the correction
is not too serious except in the case of Al.

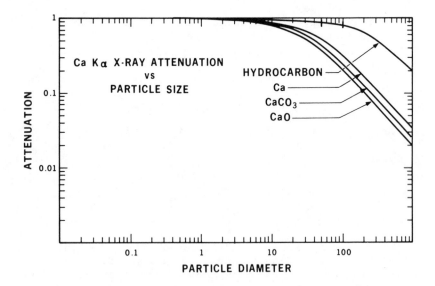

Figure 7.12. Calcium X-ray attenuation vs. particle size.

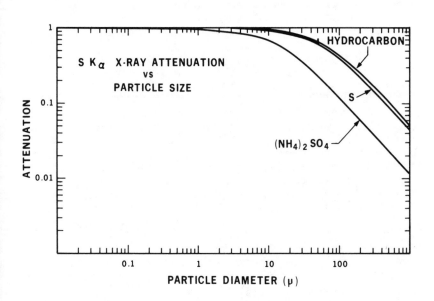

Figure 7.13.   *Sulfur X-ray attenuation vs. particle size.*

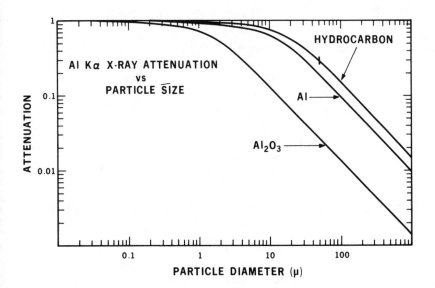

Figure 7.14.   *Aluminum X-ray attenuation vs. particle size.*

These families of curves represent a guess as to likely chemical constituents of particles. It is possible that more difficult combinations of elements might produce significant absorption effects (PbS is an obvious candidate). Again we are faced with the necessity of making some assumptions regarding the nature of particulates in order to estimate the correction factor.

The problems associated with these effects are of course inherent to the X-ray fluorescence method and are the same regardless of how one excites or detects the radiation. However, additional information can be obtained by using a monoenergetic X-ray source to excite the characteristic radiation. As noted in Equation (1), the correction factor depends upon the absorption coefficient for both the incident and emitted X-rays. By varying the incident X-ray energy, two measurements can be performed, one in which absorption of the incident radiation is negligible over the particle diameter, and the other where it is significant. Another way of looking at the problem is to consider the higher energy excitation as a probe measuring the total particle volume, whereas the low-energy excitation samples the surface only. In this way information regarding the absorption characteristics of the particle can be obtained. To a first approximation, this measured absorption correction would be independent of any assumptions regarding particle shape or composition. A similar argument could be applied to the question of matrix absorption within the filter.

SUMMARY AND CONCLUSIONS

The successful application of recently developed energy dispersive X-ray analysis techniques to the automated elemental analysis of atmospheric aerosol particulates has been demonstrated. Detectabilities for a number of environmentally important elements have been shown to be within practical limits for research and monitoring purposes. Although the emphasis of the design has been toward optimum results for air filter samples, the basic ideas can also be applied to the multielement analysis of a large class of samples.[10] Future improvements in the system performance may be realized by the incorporation of high-rate pulse excitation[11] and polarization scattering geometries.[12] Current work is being

devoted to those and other innovations that might
improve the calibration of the system.

ACKNOWLEDGMENT

We wish to acknowledge the important contribution to the
design of the system made by D. Landis and B. Loo. We also
acknowledge the efforts of F. Gin, A. Jue, W. Searles, G.
Skipper and S. Wright in the fabrication and testing of the
equipment; J. Walton and H. Sommer in detector fabrication;
and J. Anderson and D. Malone in the construction of mechanical
assemblies. We have profited from discussions and consultations
with many people including T. Dzubay, R. Hammerle and R. Giauque.
Further appreciation is expressed to R. Giauque for his continuing cooperation in the analysis of samples for comparison
purposes.

REFERENCES

1. Goulding, F. S. and J. M. Jaklevic. "X-Ray Fluorescence
    Spectrometer for Airborne Particulate Monitoring," Final
    Report to the Environmental Protection Agency, EPA Report
    No. EPA-R2-73-182 (1973).
2. Liebhafsky, H. A., H. G. Pfeiffer, E. H. Winslow, and
    P. D. Zemany. *X-Ray Absorption and Emission in Analytical
    Chemistry* (New York: John Wiley and Sons, 1960).
3. Birks, L. S. *X-Ray Spectrochemical Analysis* (New York:
    Interscience Publishers, John Wiley and Sons, 1969).
4. Giauque, R. D., F. S. Goulding, J. M. Jaklevic, and R. H.
    Pehl. "Trace Element Determination with Semiconductor
    Detector X-Ray Spectrometers," *Anal. Chem., 45,* 671 (1973).
5. Russ, J. C. *Energy Dispersion X-Ray Analysis: X-Ray
    and Electron Probe Analysis* (Philadelphia: American
    Society for Testing and Materials, ASTM Publ. 485, 1970).
6. Cooper, J. A. *Nucl. Instrum. Methods, 106,* 525 (1973).
7. Cahill, T. A. "Cyclotron Analysis of Atmospheric Con-
    taminents," Final Report to the California Air Resources
    Board, Crocker Nuclear Lab Report No. UCD-CNL-162 (1972).
8. Goulding, F. S., J. M. Jaklevic, B. V. Jarrett, and D. A.
    Landis. "Detector Background and Sensitivity of X-Ray
    Fluorescence Spectrometers," *Advances in X-Ray Analysis,
    15,* 470 (New York: Plenum Press, 1972).
9. Hammerle, R. H., R. H. Marsh, K. Rengan, R. D. Giauque,
    and J. M. Jaklevic. "A Test of X-Ray Fluorescence as a
    Method for Analysis of the Elemental Composition of
    Atmospheric Aerosol," *Anal. Chem., 45, 1939 (1973).*

10. Jaklevic, J. M. and F. S. Goulding. "Semiconductor
     Detector X-Ray Fluorescence Spectrometry Applied to
     Environmental and Biological Analysis," *IEEE Trans. Nucl.
     Sci. NS-19(3),* 384 (1972).
11. Jaklevic, J. M., F. S. Goulding, and D. A. Landis. "High
     Rate X-Ray Fluorescence Analysis by Pulsed Excitation,"
     IEEE Trans. Nucl. Sci. NS-19(3), 392 (1972).
12. Dzubay, T. G., B. V. Jarrett, and J. M. Jaklevic. "Back-
     ground Reduction in X-Ray Fluorescence Spectra Using
     Polarization," *Nucl. Instr. Methods, 115(1), 297 (1974).*

CHAPTER 8

AN INSTRUMENT FOR CONTINUOUS MONITORING OF NITRATE IN ATMOSPHERIC PARTICULATE

John N. Driscoll and Larry J. Forney

INTRODUCTION

Emissions of oxides of nitrogen from combustion processes total over 20 million tons annually in the United States alone.[1] Nitric oxide is the most stable oxide in the exhaust gases; however, it is oxidized in the atmosphere to NO_2. Its half life is between two and four hours. This is one of the first steps in a series of reactions producing photochemical smog. The intensity of photochemical smog increases in a rather complex manner with an increase in the nitrogen oxides concentration.[2]

Early meteorological models for photochemically active species described by Wayne, *et al.*[3] were only moderately successful in their predictions for the nitrogen oxides. More recent photochemical models[4-6] have incorporated additional reactions involving the oxides of nitrogen, which appear to increase the predictive accuracy of the models. Some of the postulated reactions are:

$$NO_2 + O_3 \rightarrow NO_3 \tag{1}$$

$$NO_2 + NO_3 \xrightarrow{H_2O} HNO_3 \tag{2}$$

$$NO_2 + OH \rightarrow HNO_3 \tag{3}$$

$$NO_3 + NO_2 \rightarrow N_2O_5 \qquad (4)$$

$$N_2O_5 + H_2O \rightarrow HNO_3 \qquad (5)$$

$$NO + NO_2 \xrightarrow{H_2O} 2HNO_2 \qquad (6)$$

$$HNO_2 \xrightarrow{h\nu} NO + OH \qquad (7)$$

Reactions 1-5 reduce the NO_2 levels and increase the accuracy of NO_2 predictions. *Note that HNO_3 aerosol is the product in these reactions.* Reactions 6 and 7 produce hydroxyl radicals that could oxidize additional NO via Reaction 3.

In addition, high particulate nitrate levels have resulted in failures of relay springs used by the telephone companies. When NO_3^- concentrations exceed 2-4 $\mu g/cm^2$ and the relative humidity is greater than 50%, failures take place.[1] New York City scientists have shown that unfavorable weather conditions and NO_2 produced during dynamite blastings resulted in acid aerosols that damaged nylons.[1]

Most of the present data on particulate nitrate levels is collected with a Hi-volume sampler over a 24-hour period and returned to the laboratory for wet chemical analysis. This approach does not provide the necessary resolution over the 24-hour period needed to study the role of nitrate aerosols in photochemical smog reactions, health effects, episode prediction, etc. A more rapid technique is clearly necessary.

Our objective was to design and fabricate a nitrate aerosol monitor for atmospheric measurements that would provide a minimum resolution of several hours for a nitrate aerosol concentration of 1 $\mu g/m^3$. The device had to be capable of performing three functions:

1. quantitative extraction (collection) of particulate from the atmosphere
2. conversion of particulate nitrate to a soluble form
3. analysis of the nitrate solubilized.

Our approach involved coupling two commercially-available components to achieve the above functions.

A LEAP sampler, which is, in essence, a wet electrostatic precipitator, performs the first two while a nitrate electrode provides the most direct approach to nitrate analysis.

A description of the laboratory feasibility studies of each component taken separately and then combined provides the basis for the design of the nitrate aerosol monitor. These results and the tests on atmospheric samples are discussed in the following sections.

COLLECTION TECHNIQUE

Atmospheric particulate matter traditionally has been collected on 8 x 10-in high-efficiency glass fiber filters with Hi-volume samplers operated for 24 hours at 20-50 cubic feet per minute (CFM). The nitrate on the filter is extracted in the laboratory and analyzed by a wet chemical method.* This technique is time-consuming and difficult to automate. A tape stain sampler could be considered an automated version of collection by filtration; however, the air flow through these instruments is so low (about several cubic feet per hour) that a collection time greater than 10 hours would be required. Impaction is another technique utilized for collection of particles but the small size of nitrate aerosols (MMD = 0.23, 0.59μ) found by Lee and Patterson[7] clearly eliminates the usefulness of this technique.

The technique that appeared to meet the qualifications outlined in the introduction was the wet electrostatic precipitator or LEAP sampler (Environmental Research Corp. Model 3440). This device was developed about ten years ago for concentrating biological species. The LEAP sampler concentrates several hundred liters of air containing atmospheric nitrate particulate into a few milliliters of liquid. The device utilizes inertial and electrostatic forces to impact the suspended nitrate particles onto a moving liquid film (water). The particulate nitrate dissolves, and the NO_3^- solution to be analyzed is pumped from the sampler. A schematic of the LEAP sampler is shown in Figure 8.1. The range of operating conditions specified by the manufacturer **is** as follows: a variable air flow rate of 300-1200

---

*Additional information on this subject is given in the following section.

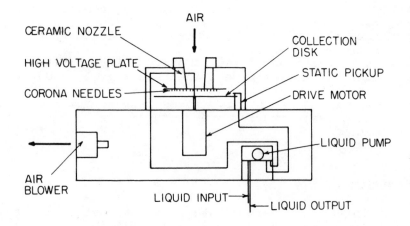

Figure 8.1. Schematic of LEAP sampler (after ERC Catalog 1969-1970).

liters per minute, a variable liquid flow rate of 0-10 ml per minute, and an adjustable precipitator voltage of 0-20 kV.

Laboratory tests demonstrated that the maximum voltage setting was about 16 kV due to arcing, and the minimum liquid flow rate was 3-4 ml/minute because of evaporation of water from the spinning collector plate. Another limiting parameter found was the air flow rate. The LEAP sampler can be operated at flow rates up to 1200 LPM, but the efficiency for collection of small particles drops sharply as indicated in Figure 8.2. Since the nitrate particles in the air are small (MMD = 0.23, 0.59μ),[7] a maximum flow rate of 600 LPM was chosen. This provides a collection efficiency of 85% even for very small particles.

The LEAP sampler can be operated in two modes: a continuous mode where the collection fluid is pumped at a specified rate for a single pass through the system, and a recirculation mode whereby a known volume of fluid is recirculated through the sampler. If ideal aerosol collection is assumed, the suspended atmospheric nitrate concentration is

$$C = C_s \frac{V}{V_a} \qquad (8)$$

where

C = atmospheric nitrate concentration
C_s = solution nitrate concentration
V = solution volume
V_a = sampled atmospheric volume.

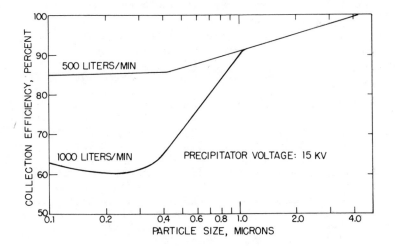

Figure 8.2. *LEAP collection efficiency (after ERC Catalog*
1969-1970).

In the continuous mode, Equation (8) can be rewritten
in terms of flow rates as follows:

$$C = C_s q/Q \tag{9}$$

where

Q = air sample rate
q = solution flow rate

However for the recirculation mode, Equation (8) can
be rewritten in terms of the air sampling rate and
sampling time as follows:

$$\bar{C} = C_s \frac{V}{Qt} \tag{10}$$

where

$\bar{C}$ = time average atmospheric NO_3^- concentration

Q = air sample rate
t = sample time
V = recirculation volume

DETECTION

A brief review of colorimetric procedures for
nitrate detection indicated that the number and
complexity of steps involved were too complex for
our purposes. The nitrite methods, although easily
automated, required a reduction step prior to
analysis. We chose the liquid ion exchange nitrate
electrode for detection of NO_3^- since it measured
nitrate directly in aqueous solution, had a wide
dynamic range (10^{-5} to 10^{-2}M), and could be easily
interfaced with the LEAP sampler described previously
The mechanism of operation for the electrode has
been described in detail by Ross.[8] Briefly, it
consists of a measurement of the potential developed
across a thin film of an organic liquid, which contains the ion exchanger. The ion exchange solution
is separated from the measurement solution by a membrane. The electrode responds to nitrate ions
according to the Nernst equation:

$$E = E_o - 2.3 \frac{n\ RT}{F} \log (NO_3^-) \tag{11}$$

where

E = potential measured
E_O = potential due to reference electrode
$2.3 \frac{n\ RT}{F}$ = Nernst constant ($\sim$ 59 MV/decade)
NO_3^- = nitrate concentration

The observed electrode response for our system
was Nernstian (*ca.* 55 MV/decade) from about 4×10^{-5}M
to 10^{-1}M NO_3^-.
The liquid ion exchanger is only partially selective for NO_3^- since the electrode responds to a number
of other dissolved species as well as NO_3^-. These
are indicated in Table 8.1 where selectivity
constant K_x for NO_3^- = 1.0. It is fortunate that
two of the most serious interferences, ClO_4^- and
ClO_3^-, are not present in atmospheric particulate.
From the data in Table 8.1, it appeared that
Br^- could be a serious interference, particularly

Table 8.1

Potential Interferences for the Nitrate
Electrode Method of Analysis

Species	Selectivity Constant[8] (K_x)	Present in Atmospheric Particulate	
		Yes	No
ClO_4^-	10^3		X
I^-	20	X	
ClO_3^-	2		X
Br^-	0.13	X	
HS^-	0.04	X	
NO_2^-	0.04	X	
CN^-	0.01		X
HCO_3^-	9×10^{-3}	X	
Cl^-	4×10^{-3}	X	
OAc^-	4×10^{-4}		X
$CO_3^=$	2×10^{-4}	X	

for suspended particulate samples collected near a highway where Br^--containing particulate from automotive emissions is present. Studies[9] carried out near Route 2 in Leominster, Mass. demonstrated that an interference ranging from 3-12% was observed due to the high Br^-/NO_3^- ratios.

One method studied for removal of halide interferences was a silver ion exchange resin.[10] In our system, we utilized a fluoride electrode for the reference;[11] therefore 10^{-2}M silver fluoride was added to the solution to eliminate the halide interference and provide a source of fluoride for the reference electrode. The 10^{-2}M silver ion eliminates the bromide, iodide and sulfide since the equilibrium levels are reduced to 5×10^{-11}, 8×10^{-15}, and 10^{-50} respectively. The effect of bromide on the nitrate electrode with silver fluoride addition is shown in Table 8.2. No difference in the readings was observed, even for 10^{-5}M NO_3^- where a 40-fold Br^-/NO_3^- ratio occurs. These laboratory data indicated that a potentially interference-free method for nitrate in atmospheric particulate has been

Table 8.2

Effect of Bromide on Nitrate Electrode Using
Silver Fluoride Electrolyte

(NO_3^-)	MV*	MV**
Mud	+190	+190
10^{-5}M	+184	+183
10^{-4}M	+144	+145
10^{-3}M	+ 87	+ 87
10^{-2}M	+ 29	+ 29
10^{-1}M	- 27	- 28

*25 ml of standard + 1 ml 10^{-1}M AgF + 1 ml deionized H_2O
**25 ml of standard + 1 ml 10^{-1}M AgF + 1 ml 10^{-2}M Br⁻

developed but further confirmation by actual samples
was required.

Atmospheric samples were collected on 8 x 10-in
glass fiber filters with Hi-volume samplers.   The
filter papers were sectioned and leached with water,
and the aqueous aliquots were analyzed by three
different wet chemical methods.   The results are
given in Table 8.3.   The Sawicki method[12] involves
reduction of nitrate to nitrite by hydrazine catalyzed
by cupric ions.   The nitrous acid found is analyzed
via a diazotization-coupling reaction with a spec-
trophotometric readout.   In the PDS method,[13] samples
are evaporated to dryness, and phenol in fuming
sulfuric acid is nitrated.   The colored product,
2,4-dinitrophenoldisulfonic acid, is determined
spectrophotometrically at 400 nm.

The correlation coefficients between the methods
are quite high, the slopes are near unity, and inter-
cepts are small as seen in Table 8.4.

In view of the above data, we have demonstrated
that the detection of nitrate by the electrode method
yields results comparable to conventional wet chemical
methods.

Table *8.3*

Comparison of Nitrate in Atmospheric Particulate
By Colorimetric and Nitrate Electrode
Analyses[9]

Sample No.	NO_3^- Level ($\mu g/7.5\ in^2$)		
	Nitrate Electrode	PDS	Sawicki
11	566	324	342
12	217	114	194
13	264	224	256
14	174	150	160
15	3007	2824	2800
16	488	414	477
17	245	220	232
18	484	524	543
19	264	254	221
20	20	68	29

Table *8.4*

Statistical Analyses of Nitrate Method Data

Method	Correlation Coefficient	Regression Analysis (y = electrode, x=colorimetric)
Electrode *vs.* Sawicki	0.896	y = 0.989x + 33
Electrode *vs.* PDS	0.869	y = 1.05x + 36
Sawicki *vs.* PDS	0.971	y = 1.06x + 3

INSTRUMENT DESIGN

The nitrate monitor consists of two basic units:
the LEAP sampler and a flow-through electrode sensing
cell, with associated electronics. A schematic of
the system is shown in Figure 8.3. The nitrate-laden
solution is pumped from the LEAP sampler through a
temperature controlled lucite sensing cell, which
holds the nitrate electrode (Orion Model 92-07),
fluoride reference electrode (Orion Model 94-09),

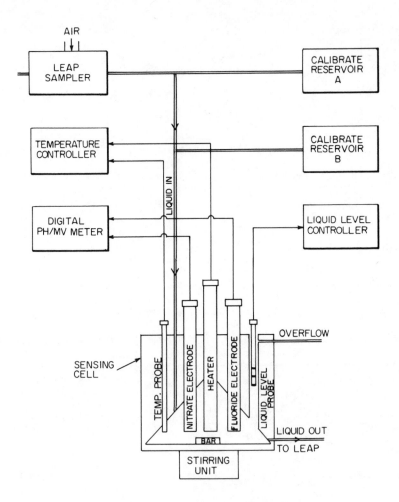

Figure 8.3. Schematic of nitrate monitor.

a temperature probe (Versa Therm Model 8446), an
immersion heater (Pyrex Cat. No. 33847-026), a
stirring bar, and three liquid level sensing probes
(Dyna-Sense Cat. No. 7186-12). The digital
millivoltmeter (Orion Model 701) displays the
output from the electrodes.

The sensing cell, in addition to housing the
electrodes, maintains the collection solution at a
constant temperature above ambient with the immersion heater (the electrode response is temperature-
dependent), stirs the solution to smooth inhomogenei-
ties and provides a uniform flow past the electrodes,
and provides a constant liquid level to compensate
for evaporative losses during the recirculation
mode. The details of the nitrate detector design
are discussed by McCoy, *et al.*[14]

In addition to the continuous and recirculation
modes outlined earlier, the present system provides
for either automatic or manual calibration and flush
cycles.

During laboratory tests, convenient operating
conditions for the LEAP sampler were found to be:
an air flow rate of 600 liters/min, a liquid flow
rate of $\sim$ 10 ml/min, a precipitation voltage of
14.5 kV, and a sensing cell temperature of 38.5°C.
The low air flow rate increased the efficiency of
the LEAP sampler for small particles < 1μ, while
a constant cell temperature of 38.5°C was chosen
in order to be well above ambient levels. The fully
equipped cell volume was 44 ml, which yields a cell
time constant $\tau \sim 4$ min. The recirculation volume
was measured (cell volume plus fluid in tubing and
sampler) and found to be 60 ml. Thus, for an air
flow rate of 600 liters/min, an atmospheric nitrate
concentration of 1 μg/m³ yields an incremental
increase in NO_3^- concentration of 1×10^{-5} M for a
1-hour sampling period.

Since atmospheric nitrate levels are typically
in the range of 1-3 μg/m³ as indicated in Table
8.5, the nitrate and fluoride electrodes will be
in the *non-Nernstian* region in the first 1 to 4
hours using the recirculation mode. The instrument
can be operated in the Nernstian region initially
by starting the sampling period with a recirculation
collection fluid spiked with 4×10^{-5} M NO_3^-. Now,
as atmospheric nitrate is scrubbed into the solution,
the nitrate level will increase. For an electrode
with a 58 MV/decade slope, an 18 MV change will
correspond to an increase in nitrate concentration
of 4×10^{-5} M. This is similar to the known addition

Table 8.5

Atmospheric Nitrate Aerosol Concentrations [9]

Site	Concentration Range (μg/m³)	Normal Concentration (μg/m³)
Urban	0.5-20	3
Nonurban	0.1-8	0.8

technique described by Orion Research.[15] We found
that the range of operating parameters of the system
precludes the use of the continuous mode except for
cases of unusually high atmospheric nitrate levels
(> 10 μg/m³).

PRELIMINARY ATMOSPHERIC TESTS

Two 4.5-hour atmospheric tests were conducted
with the nitrate monitor in the recirculation mode.
The results are shown in Figure 8.4.   The contours
were drawn in Figure 8.4 by assuming that the suspended nitrate concentration was constant in time
and that the nitrate electrode was operating in the
Nernstian region (the latter would not be the case
for low atmospheric NO_3^- levels and/or short sample
times). With these assumptions, Equations (10) and
(11) can be combined to give:

$$\frac{\Delta E - \Delta E_o}{K} = -\log\left(\frac{\bar{c}\,Qt}{[F^-]\,V}\right) \tag{12}$$

where the standard potential ΔE_o and slope K were
determined by two calibration points for the system
taken prior to each test. From Figure 8.4, the
assumptions leading to Equation (12) appear to be
correct during the last two hours of each test and,
in fact, a time-averaged nitrate concentration of
∿ 2 μg/m³ is indicated for both tests.

The voltage response of the system is shown in
Table 8.6, along with average atmospheric NO_3^-
concentrations for 30-minute increments during the
last 2 hours of each test.

To determine the efficiency of collection of the
nitrate detector system and to compare the detector

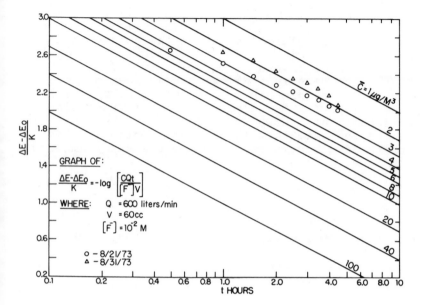

Figure 8.4. *Average suspended NO_3^-.*

Table 8.6

Voltage Response and Average Atmospheric NO_3^- Concentrations

Time (hr)	Run 1 8/21/73 (MV)	$\overline{c}$ (30min avg) ($\mu g/m^3$)	Run 2 8/31/73 (MV)	$\overline{c}$ (30min avg) ($\mu g/m^3$)
0	248		238	
1/2	236			
1	229		231	
1-1/2	220		224	
2	214		219	
2-1/2	210		214	
3	207	1.2	211	1.0
3-1/2	204	1.6	208	1.2
4	200	2.4	203	2.4
4-1/2	197	2.0	197	3.4

with the conventional measuring technique, *e.g.*, a
high volume sampler, an 8-hour test was run on
9/18/73 from 1018 to 1818. During this test, the
nitrate detector was run simultaneously with three
high-volume samplers located at the apices of an
isosceles triangle 15 feet on a side. The high-
volume sampler inlets were facing away from each
other along each leg, all within 15 feet of the
detector. The nitrate was leached from the high-
volume filters and analyzed using a nitrate electrode.
The results are shown in Table 8.7. These preliminar
results indicate that the nitrate monitor yields
nitrate levels comparable to the standard sampling
technique.

Table 8.7

*Comparison of Nitrate Monitor and Hi-Volume Sampler Results,
NO$_3^-$ Levels (µg/m^3)*

Nitrate Detector	*Hi-Vol Samplers*
1.5	1.8, 1.1, 1.7
	1.53 avg.

ACKNOWLEDGMENT

The work described in this chapter was performed by the
Walden Research Division of Abcor, Inc. pursuant to EPA
Contract No. 68-02-0591.

REFERENCES

1. "Air Quality Criteria for Nitrogen Oxides," EPA Report
   AP-84 (1970).
2. Khan, A. W., J. N. Pitts, and E. B. Smith. *Environ. Sci.
   Technol.*, *1*, 656 (1967).
3. Wayne, L. G., *et al.* "Modeling Photochemical Smog on a
   Computer for Decision-Making," presented at the 63rd
   Annual APCA Meeting (1970).
4. Seinfeld, J. H., *et al.* "Simulation Model for Estimation o:
   of Ground Level Concentrations of Photochemical Pollutants,"
   Systems Application Corp., sponsor EPA, NTIS No. PB 206-410
   (1973).

5. Wayne, L. G., *et al.* "Evaluation of the Reactive Environmental Simulation Model," Pacific Environmental Services, sponsored by EPA, NTIS No. PB 220-456 (1973).

6. Martinez, J. R. and A. Q. Eschenroder. "Further Development of a Smog Model for Los Angeles Basin," General Research Corp., sponsored by EPA, NTIS No. PB 201-737 (1972).

7. Lee, R. E. and R. K. Patterson. "Size Determination of Atmospheric Phosphate, Nitrate, Chloride, and Ammonium Particulate in Several Urban Areas," *Atmos. Environ., 3*, 249 (1969).

8. Ross, J. W. in *Ion Selective Electrodes*, R. A. Durst, ed. NBS Special Publication No. 314 (1969).

9. Williams, D., J. Driscoll, C. Curtin, R. Hebert. "Methods for the Rapid and Accurate Measurement of Nitrate and Sulfate in Atmospheric Particulate," Walden Research Div. of Abcor, EPA Contract No. 68-02-0564 (December 1973).

10. Paul, J. and J. Carlson. *Agr. Food Chem., 16,* 766 (1968).

11. Manahan, S. E. *Anal. Chem., 42,* 128 (1970).

12. Sawicki, C. R. and F. P. Scaringelli. *Microchem. J., 16,* 657 (1971).

13. *ASTM Annual Book of Standards*, Part 23, method D1608 (Philadelphia, Pennsylvania: ASTM, 1971).

14. McCoy, J. F., *et al.* "Development of a Prototype Nitrate Monitor," Walden Research Div. of Abcor, EPA Contract 68-02-0591 (1974).

15. "Instruction Manual Nitrate Ion Electrode," model 92-07, Orion Research Corp. (1971).

CHAPTER 9

APPLICATION OF AN OSCILLATING QUARTZ CRYSTAL TO
MEASURE THE MASS OF SUSPENDED PARTICULATE MATTER

Raymond L. Chuan

DESCRIPTION OF BASIC TECHNIQUE

The quartz crystal microbalance particulate mass
monitor (QCM/PM) is basically an active impactor in
which particulates are impacted by aerodynamic
acceleration in a conventional way against a thin
quartz disc (about 1.2 cm in diameter and 0.015 cm
in thickness) which is part of a high frequency
oscillator circuit whose frequency (nominally 10 MHz)
is controlled by the crystal. Addition of mass
through impaction and capture by a thin adhesive
layer on the crystal causes a decrease in its
resonant frequency, the change being proportional
to mass change over a wide dynamic range of about
10^5. Placed in close proximity to the sensing
crystal, but not receiving any impacted mass, is an
identical reference crystal controlling the frequency
of another circuit at a slightly higher frequency
than that of the sensing oscillator. The beat-
frequency between the two oscillators (of the order
of 2 KHz) is thus the signal reacting to mass change.
The slight temperature sensitivity of the crystal is
compensated for by the use of the reference crystal.
The basic sensitivity of such a device is of the
order of 10^9 Hz/gm, making it possible to detect a
mass change as little as 10^{-11} gm.
By maintaining a constant volume rate of sampling
flow to the sensing crystal the mass concentration
of particulates is proportional to the rate of change
of frequency of the crystal. The beat-frequency is

163

converted to a voltage and then differentiated with respect to time, so that the output from the differentiator is scaled to read directly in particulate mass concentration. The time constant of the differentiator can be selected for a wide range of values, from less than 1 second to 10 seconds. Electronic and flow schematics of a representative instrument are shown in Figure 9.1.

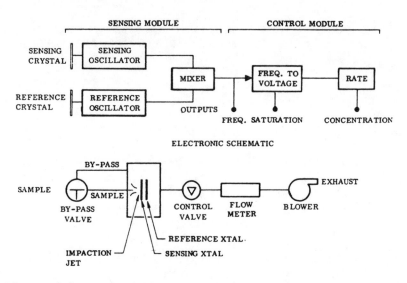

Figure .9.1.   Sample air flow and electrical schematic.

If the number density of particulates is small enough such that impactions at the crystal are not coincident, it is possible to detect discrete impacts with the time constant at less than 0.01 second. The detection limit in this case is 10^{-11} gm, or the equivalent of a 2-micron diameter density 2 gm/cm^3 particle. With the short time constant, the discrete impact of a particulate mass results in a pulse in the differentiator output whose height is the mass of the particle. This is shown in Figure 9.2a If the particulate matter is volatile, so that it evaporates after impaction, the signal pulse is followed by a negative pulse—the result of mass loss and attendant cooling of the crystal. (The frequency-temperature behavior of the crystal is chosen so that heating results in frequency increase.) Figure 9.2a thus shows the signal from nonvolatile particulates, while Figure 9.2b shows volatile particulates.

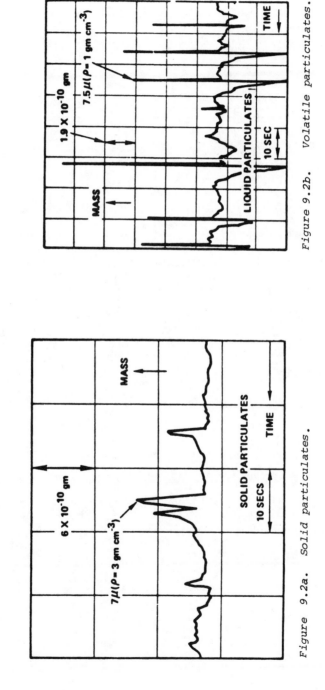

Figure 9.2b. Volatile particulates.

Figure 9.2a. Solid particulates.

Smaller masses are measured cumulatively, over time intervals of 0.1-10 seconds. An example is shown in Figure 9.3, where one trace is the cumulated mass and the other is the mass concentration.

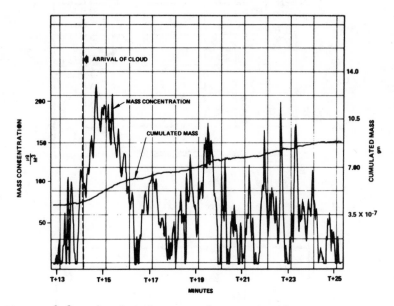

Figure 9.3. Cumulated mass and concentration.

SYSTEM SENSITIVITY AND CALIBRATION

The sensitivity of the quartz crystal is related basically to its thickness or its resonant frequency. As determined by Sauerbray,[1] the theoretical sensitivity of a piezo-electric crystal oscillating in the thickness shear mode is

$$\frac{\Delta f}{\Delta m/A} = 2.2 \times 10^{-6} \; f_o^2 \quad \frac{Hz \; cm^2}{gm}$$

where f_o is the resonant frequency of the crystal and A is the active area of the crystal which, in practice, is equivalent to the area of the electrodes.

For a 10 MHz crystal with electrode area of 0.385 cm^2 the sensitivity works out to be

$$\sigma = 5.72 \times 10^8 \text{ Hz/gm}$$

If C is the mass concentration of particulates (in μgm^{-3}) and V is the volume rate at which the particulate-laden gas is directed at the crystal by the impaction jet (in ml min^{-1}), the frequency shift rate of the sensing crystal is

$$\frac{\Delta f}{\Delta t} = 10^{-12} \sigma C V \text{ Hz min}^{-1}$$

In most of the QCM instruments tested so far a total sampling flow rate of 1500 ml min^{-1} is used, with only part of this directed at the crystal, the remainder being by-passed around the crystal. This is done to provide a reasonable flow velocity in the sampling line to minimize settling of the larger particles. Usually four impaction jets each 0.04 cm in diameter are used, with an aggregate volume rate of V = 600 ml min^{-1}. With such an impaction arrangement and a particulate mass concentration of 50 μgm^{-3}, the response rate of the instrument is

$$\frac{\Delta f}{\Delta t} = 17.2 \text{ Hz min}^{-1}$$

With the circuitry developed so far, a frequency shift rate down to 1 Hz min^{-1} can be measured, which would allow one to measure concentration down to 3 μgm^{-3}. Of course, this and lower concentrations can be effectively measured by increasing the sampling flow rate, without loss of impaction efficiency, by using a large number of small impaction jets.

The QCM instrument has been calibrated by a number of methods: direct system calibration with measured quantities of polydisperse particulates in a fixed air volume, scanning-electron-microscope analysis of captured particles, comparison with sampling train measurement, and direct comparison with high-volume filtration samplers. In this last method the sensing head of the QCM was placed directly above the filter holder. Sampling air flow through the sensor head was turned on for 1 minute every 10 minutes during the 3.5 hours test period. Table 9.1 shows the results.

Table 9.1

Comparison of QCM and Hi-Vol

Time	Freq. Rate* Hz/min $\dot{f}$	Freq.** Hz f	Mass Conc. $\mu g/m^3$
1340 Start Hi-vol (Initial filter weight = 3.007 gm. Flow rate = 69 ft^3 min^{-1})			
41-42	60		187
51-52	60	1668	187
1401-02	80	1724	250
11-12	45	1797	141
21-22	50	1856	156
31-32	40	1916	125
41-42	40-80	1987	125-250
51-52	40-90	2054	125-281
1501-02	60	2102	187
11-12	40	2165	125
21-22	40	2214	125
31-32	50	2264	156
41-42	60	2314	187
51-52	40	2375	125
1601-02	35	2422	109
11-12	35	2463	109
21-22	30	2501	91
31=32 Stop Hi-vol	35	2537	109

(Final filter weight = 3.062 gm)

*Frequency rate read 40 seconds after beginning of 1 minute sampling period.
**Frequency read at end of 1 minute sampling period.

Average mass concentration by Hi-Vol	= 164 $\mu g/m^3$
Total mass impacted on crystals	= 1.52 x 10^{-6} gm
Total volume sampled by QCM	= 9.53 x 10^{-3} m^3
Average mass concentration sampled by QCM	= 160 $\mu g/m^3$
Average of instantaneous concentration readings from QCM	= 154 $\mu g/m^3$

POST-SAMPLING ANALYSIS

Since the adhesive-coated crystal retains the impacted material, post-sampling analysis can be performed; because the undisturbed crystal can be placed directly in a SEM, such analysis can be performed nondestructively and without disturbing the material by the use of scanning-electron-microscope (SEM). Figure 9.4 shows a 100 x SEM of an entire impaction area (taken at normal incidence), measuring about 500 microns across, produced by exposing the crystal through a 400-micron impaction nozzle to smoggy atmosphere for about 10 minutes. The measured concentration was 250 μgm^{-3}. At an impaction flow rate of 150 ml/min^{-1}, this means a total mass accumulation on the crystal of about 0.38 μg. The symmetrical distribution of material, with smaller particles spread to the outer regions, is quite apparent.

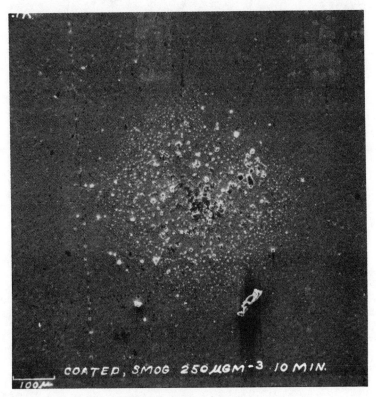

Figure 9.4.   100 x SEM of impaction area.

Figure 9.5 is a 10,000 x SEM (at 10 degrees tilt) of a part of the impaction area shown in Figure 9.4. A wide range of particle sizes is discernible, from somewhat less than 0.1μ to about

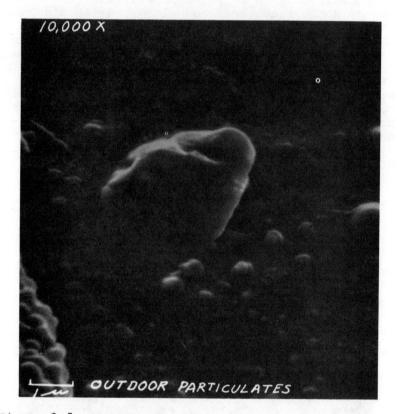

Figure 9.5. 10,000 x SEM.

3μ. In the SEM, X-rays scattered from the material being studied can be analyzed to yield elemental composition and, further, to identify molecular composition from relative molal intensities of the elements. By scanning the material field at a fixed X-ray energy (related to a particular element) a map of that element can be established.

Figure 9.6a shows a 10,000 x SEM of material captured on a crystal during a sampling flight through the plume of a solid-fueled rocket. The spherical particles seen are in this case aluminum oxide. An aluminum X-ray map of the same field is

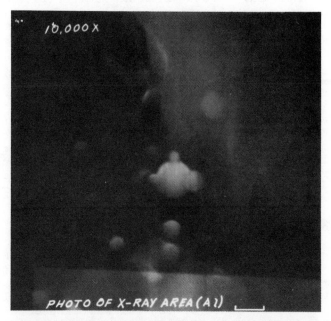

Figure 9.6a.   10,000 x SEM of Al₂O₃.

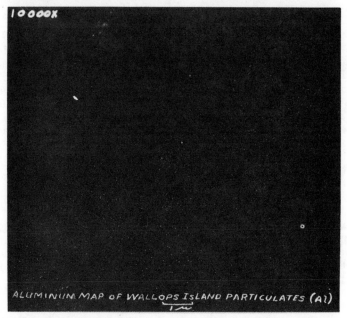

Figure 9.6b.   X-ray map.

shown in Figure 9.6b, confirming that the spheres
contain aluminum. It is worth noting that elemental
or molecular identification can be made from scanning
a single particle less than 1μ in size and no more
than 10^{-12} gm in mass.

EXAMPLES OF APPLICATIONS

The QCM particulate mass instrument has been
used in a significant number of applications over
the past two years, in both ambient air and source
situations. A few representative examples, which
illustrate the particular advantages of the device
to the situations in question, are presented here.

Automobile Exhaust

Direct-connect exhaust measurements were made
at the EPA Ypsilanti Laboratory (in May 1971) with
an EPA-owned car on a dynamometer going through the
standard 23-minute LA-4 driving sequence. Sample
from the exhaust stack of the car was diluted with
dry clean nitrogen to a dilution ratio of 10:1, to
reduce the concentration of the sample going to the
quartz crystal and to dehumidify the sample to
avoid water condensation. An actual strip-chart
recording of the test is reproduced in Figure 9.7,
in which three traces are shown — the speed profile
of the LA-4 sequence, the cumulated particulate mass,
and the rate of mass (equivalent to concentration).
Examination of Figure 9.7 shows that in a warm
start about one-third of the total particulate mass
is emitted during the hard acceleration to 55 mph,
beginning just after 3 minutes, and that the emission rate is relatively constant for the rest of the
sequence, except for a very large increase at 19.5
minutes, which resulted from an unscheduled stalling
of the engine and a restart. Using 200 m^3 as the
total exhaust volume and 7.5 miles as the total distance traveled in the LA-4 sequence, the average
particulate emission per mile is calculated, using
concentration obtained with the QCM instrument, to
be 71.3 mg/mile. If the anomalous contribution
from the unscheduled restart is excluded, the average
particulate emission per mile is 37.0 mg/mile.

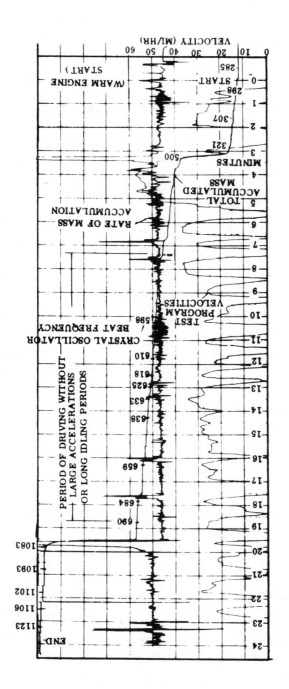

Figure 9.7. Exhaust particulate measurement.

Basic Oxygen Furnace

The emissions from a basic oxygen furnace vary considerably with the operating stage of the furnace. During charging of the furnace, for example, emissions can be quite high for a few minutes, after which they would be quite low for hours. An example of this variation in emissions is shown in Figure 9.8 which is a reproduction of the strip-chart record of the cumulated mass versus time on a QCM attached, though a 10:1 dilution (with heated dry nitrogen), to the exhaust stack of an electrostatic precipitator which cleans the particulates from the basic oxygen furnace.

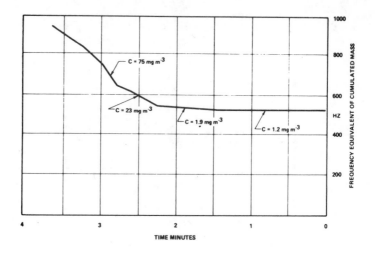

Figure 9.8. Electrostatic precipitator exhaust during BOF charging.

It is seen that for the first 1.5 minutes the rate of increase of cumulated mass is very low, this being the period just before the charging operation. The concentration during this period is 1.2 mg/m^3. At about 1.5 minutes there is slight increase in the mass rate, corresponding to a concentration of 1.9 mg/m^3, due to the preparation of the furnace just before charging. At 2.25 minutes the effect of the charging operation begins to show clearly, with a mass concentration of particulates (up to 20μ in size) of 23 mg/m^3. The concentration increases during the next 10 seconds or so to 75 mg/m^3, settling down to about 14 mg/m³ for the next 5 minutes, and finally down to about 8 mg/m³ for the next 10 minutes (the record for the last two segments not shown).

Coal-Fired Power Plant Plume

The smoke plume from the coal-fired power plant at Four-Corners, New Mexico, while of questionable environmental standing, provides an almost ideal set-up for the experimental study of dispersion because it closely approximates a fixed point source of essentially constant strength dispersing into a wide region of unobstructed and largely unpopulated landscape. By installing a QCM particulate instrument in a light plane we have conducted a number of *in-situ*, real-time measurements of the plume during the summers of 1971 and 1972.

The results to be summarized briefly here were obtained in August, 1972, over the San Juan River Basin downwind of the power plant. The ground tracks of two representative flights, one along the approximate center of the plume and the other across the plume, are shown on a map of the region in Figure 9.9.

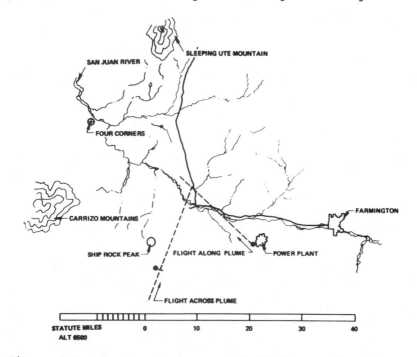

Figure 9.9.. San Juan Basin map with flight ground tracks.

The particulate mass concentration in the longitudinal flight at an altitude of 1,200 ft above ground (900 ft above the stacks) is shown in Figure 9.10. At the time of day when these measurements were made (1142 MDT) the air was starting to become unstable, so that fumigation of the plume had taken place, accounting for the rapid decay of the concentration.

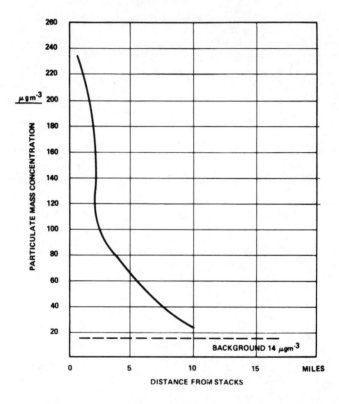

Figure 9.10. Particulate mass concentration along plume.

The transverse concentration profile of the plume taken 15 miles downwind is shown in Figure 9.11. At the time this traverse was made the two parts of the plume — one part emanating from Generator Units 1, 2 and 3 through two stacks with wet scrubbers and the other part emanating from Units 4 and 5 through two other stacks with electrostatic precipitators —

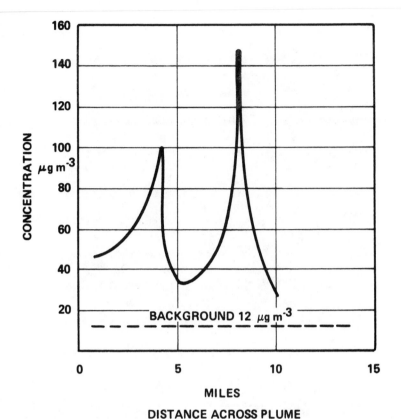

Figure 9.11. Particulate mass concentration across plume.

were still discernible. One of the four impaction
regions on the crystal used in this experiment were
studied by SEM. Figure 9.12 shows a 100x SEM of
the region, while Figures 9.13 and 9.14 are 1000x
and 2900x SEM's of the encircled part of Figure 9.12.
The largest particle seen in the field is about 10μ,
while the predominant sizes appear to be in the range
0.5-2μ. Another representative group of particles
is shown in Figure 9.15 and 9.16, where two dis-
tinct particles are seen, one jagged in shape and
the other spherical. The latter, like the many
other spherical particles seen in Figures 9.13
through 9.16, is presumably fly-ash from the stacks,
since such spherical shapes are not likely to occur
naturally.

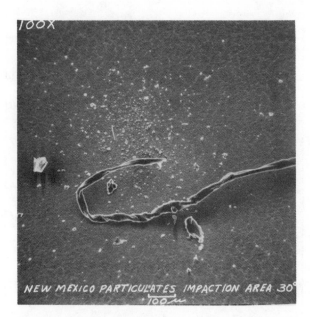

Figure   9.12.   SEM 100 x San Juan sample.

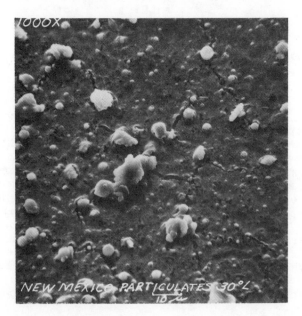

Figure   9.13.   1,000 x SEM.

Figure 9.14.   2,900 x SEM.

Figure 9.15.   1,000 x SEM of particles.

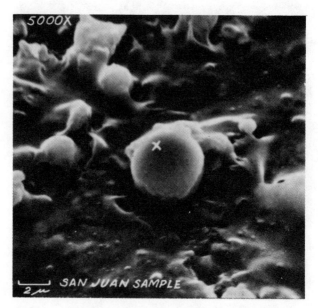

Figure 9.16.   5,000 x SEM of spherical particle.

These two particles were further analyzed by
X-ray fluorescence to determine their chemical compositions.   Two X-ray spectra are shown in Figures
9.17a and 9.17b, the former taken on the background
near the jagged particle and the latter from probing
the spot on the particle marked by an x in Figure
9.15. Palladium appears in both spectra because a
thin layer (about 60 Å) of it is evaporated over
the sample to render it electrically conducting.
Gold and silicon appear strongly in the background
because the crystal electrode is gold, and the quartz
itself contains silicon. The intensity of the
probing electron beam is kept low so as not to
penetrate beyond the sample material being analyzed.
Thus when the beam is directed at the spot marked
in Figure 9.15, only the material in the immediate
neighborhood is probed, and not the background.
From the relative intensities of the peaks in the
spectrum, the elemental composition and a probable
molecular composition can be constructed, as shown
in Table 9.2.   The probable composition of this
particle is $Ca.4K.Al_2O_3.4 Si O_2$. This presumably

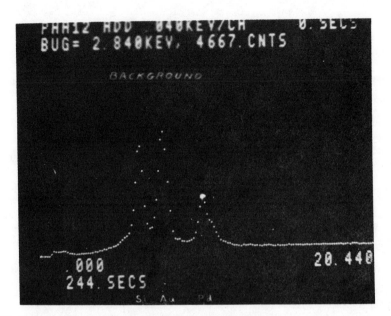

Figure 9.17a. Background X-ray spectrum.

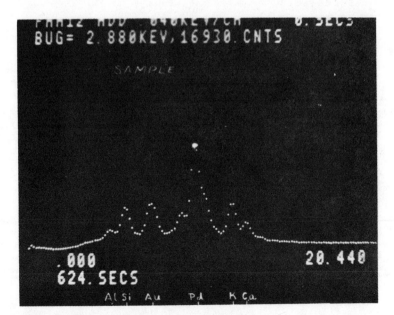

Figure 9.17b. Sample X-ray spectrum.

Table 9.2

Composition of Particle Sample

Element	Relative Intensity	Weight %	Mole %	Mole Ratio
Al	1030	14.1	17.3	2
Si	2230	30.5	36.0	4
K	3250	44.5	37.7	4
Ca	800	10.9	9.0	1

naturally occurring particle is probably a mineral of the feldspar group, possibly a mixture of ortho-clase ($K\ Al\ Si_3\ O_8$) with some other material containing calcium.

The spherical particle is probed at the spot marked by x in Figure 9.16, and the results shown in Table 9.3.   The probable composition is $8(CaO)$, $(Fe.2Mg.2Ti) \cdot 12(Al_2O_3) \cdot 40(SiO_2)$, which is considerably more complex than that of the jagged particle.   This is probably because the spherical particle is the result of high temperature processes in a furnace.

Table 9.3

Composition of Spherical Particle

Element	Relative Intensity	Weight %	Mole %	Mole Ratio
Mg	260	2.2	2.7	2
Al	3506	28.7	31.6	24
Si	5902	48.3	51.1	40
Ca	1815	14.8	11.0	8
Ti	433	3.5	2.2	2
Fe	302	2.5	1.3	1

Solid-Fueled Rocket Plume

In this application a QCM particulate instrument
was placed 3 km downwind of the launch pad of a Delta
rocket, used for launching communication satellites
at Cape Kennedy. Prelaunch dispersion calculation
had predicted arrival of the rocket cloud at T + 14
minutes. Figure 9.3, which is a reproduction of
the strip-chart recording of the cumulated particulate
mass and mass concentration obtained during the experiment, does indeed confirm the cloud arrival at
T + 14. Further, it shows a maximum concentration
of about 200 $\mu g \ m^{-3}$ in the cloud.
Figure 9.18 is a 200x SEM of the material collected on the impaction region; Figures 9.19 and
9.20 show enlargements of two of the particles
marked in Figure 9.18. Particle 3 appears to be
one of several cubical shapes, presumably salt
crystals; Particle 2, along with many similar ones
around it, appears to be a spherical combustion
product from the rocket motor. Particle 1 is quite
complicated in shape, and does not suggest anything
readily identifiable. Results of X-ray analysis of
these particles and of a reference table salt crystal
are shown in Table 9.4.

Figure 9.18. Impaction region from rocket plume 200 x SEM.

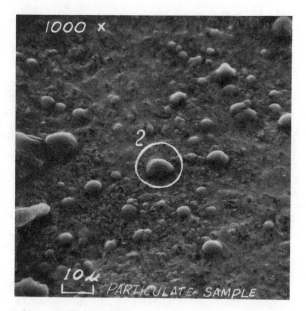

Figure 9.19.   1,000 x SEM.

Figure 9.20.   2,000 x SEM.

Table 9.4

Molal Fractions of Particle Samples

| Element | Particle | | | | |
	1-1	1-2	2	3	Table Salt
Na	34.0	23.6		50.3	41.0
Al	9.1	14.8	70.7	1.2	11.4
Si	11.0	18.0	2.9	0.8	6.5
An					
S		15.9			
Cl	38.3	5.6	26.4	47.8	41.0
K	4.7	8.6			
Ca	2.8	6.8			
Ti		6.7			

It is seen that Particle 2, with a preponderant percentage of aluminum, is probably a piece of aluminum oxide from the rocket motor. From its shape and composition it is evident that Particle 3 is salt, and a purer salt, in fact, than the reference table salt. Particle 1-1 is apparently made up of 68% salt and a few other materials. Particle 1-2, on the other hand, appears to be a different and much smaller particle adhering to the larger particle, and has a fairly complex make-up.

Vertical Distribution of Particulates
Over the Los Angeles Air Basin

A joint experiment was conducted by ourselves and NASA Langley Research Center in October, 1972, to determine the vertical distribution of particulates over the Los Angeles Air Basin by two independent methods — *in situ* QCM measurements from an airplane and ground-based laser radar. A complete report on this experiment will be forthcoming; a representative set of data is presented here. Particulate mass concentration was measured from 1,600 ft to 12,000 ft above sea-level over the San Gabriel Valley of the basin in mid-October on a day with only a light degree of apparent turbidity in the atmosphere.

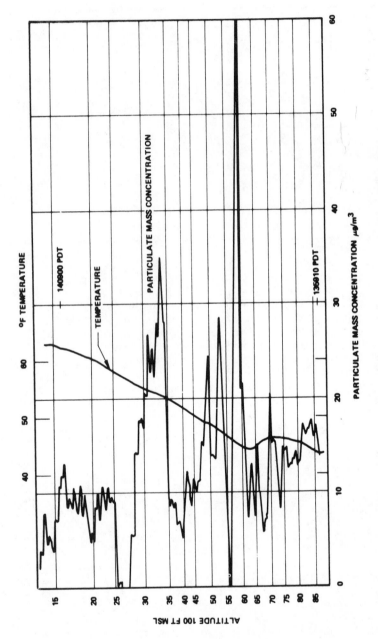

Figure 9.21. Vertical profiles of particulate concentration and temperature.

Air temperature was measured simultaneously with a thermister probe. Results of one traverse from 8,500 ft to 1,550 ft (in 10 minutes) are shown in Figure 9.21. A very dense but thin layer of particulates is seen to be trapped just below the temperature inversion at 6,000 ft.

Above the inversion the concentration decreases rapidly to about 15 $\mu g/m^3$. While in the record the the concentration trace goes out of scale, it is estimated from observation of the meter display on the instrument at the time of measurement that the peak concentration at 6,000 ft was over 300 $\mu g/m^{-3}$. Other peaks appear at 5,300 ft, 5,000 ft, 3,400 ft, and there is a thick band between 2,400 ft and 1,600 ft. It appears that these layers of particulates are related to very small changes in the temperature lapse rate. Laser radar reflection data corroborated the locations of these layers. It was noted during the experiment that over a two-day period there was no significant variation in the layer structure.

AN ACTIVE CASCADE IMPACTOR SYSTEM

A natural and obvious extension of the QCM technique is to a cascade system to obtain particulate size distribution. This has, in fact, been done in a 10-stage cascade with the characteristics shown in Table 9.5. The 50% cut-off is for particles

Table 9.5
Characteristics of Cascade Impactors

Stage	Lower 50% Cut-Off Microns	Upper 50% Cut-Off Microns	Jet Diameter mm	Number
1	47.00	100.00	2.500	1
2	22.00	47.00	2.500	1
3	10.25	22.00	2.500	1
4	4.80	10.25	1.580	1
5	2.25	4.80	0.950	1
6	1.045	2.25	0.572	1
7	0.490	1.045	0.343	1
8	0.229	0.490	0.207	1
9	0.107	0.229	0.099	2
10	0.050	0.107	0.022	37

Flow Rate = 180 ml min^{-1}

with mass density 2 gm/cm^{-3}. Figure 9.22 shows the
assembled cascade; Figure 9.23 shows the 10th
stage. Each stage has a separate QCM, with the
beat-frequency signal going through a multiplexer
to a desired data processing system. In the present
configuration the beat-frequency of each stage is
counted for 10 seconds every 2 minutes and recorded,
so that the mass fractions over the range 0.05 to
100 microns can be obtained once every two minutes,
with sufficient resolution at a total mass concentration of about 100 $\mu g/m^3$.

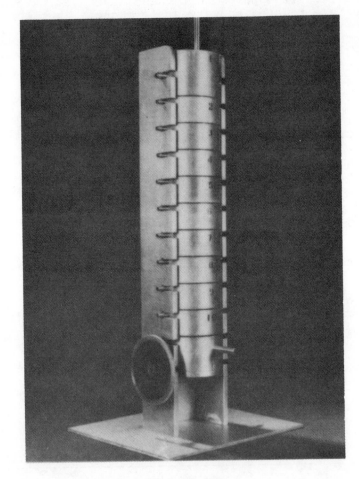

Figure 9.22. 10-stage cascade QCM.

Figure   9.23.   Details of stage 10.

REFERENCES

1.   Sauerbray, G. Z.   "Verwändung von Schwingquarzen zur
     Wägnung Dünner Schichten und zur Mikrowägung," Z. Physik,
     155, 206 (1959).

SECTION III

TECHNIQUES TO MEASURE POLLUTANTS FROM
STATIONARY AND MOBILE SOURCES

CHAPTER 10

CROSS-STACK MEASUREMENT OF POLLUTANT CONCENTRATIONS
USING GAS-CELL CORRELATION SPECTROSCOPY

Darrell E. Burch and David A. Gryvnak

INTRODUCTION

Background

Most commercially available instruments for
measuring the concentrations of gaseous pollutants
in the effluent of stationary sources require that
a sample be drawn from a stack or duct through a
probe into a separate container for analysis. The
analysis may be made continuously on-line or at a
later time, possibly in a laboratory. Such sampling
techniques have two major drawbacks: (1) concentration
changes may occur along the sampling train, and
(2) concentration gradients may exist in the flow
field so that several samples may need to be taken
from different positions in order to determine the
average concentration. By employing *in situ* absorption spectroscopy, the first difficulty can be
eliminated and the second one greatly reduced. In
absorption spectroscopy radiant energy from a thermal
source is transmitted through the effluent from one
side of the duct to the other. By comparing the
amount of transmitted energy at characteristic wavelengths where a particular gas species absorbs
strongly to that at wavelengths it does not absorb,
or absorbs weakly, it is possible to determine the
instantaneous average concentration across the flow
field. The effluent flow is undisturbed, and the
measurement along the path across the duct adequately
represents the average, unless there are unusually
large gradients in the concentration.

The required wavelength selection for absorption
spectroscopy may be achieved in a variety of ways.
Narrow band interference filters provide a simple,
relatively inexpensive method, but in most cases
they do not provide the high spectral resolution required for good specificity, or discrimination against
other gas species that may absorb in the same spectral
interval. Dispersing instruments employing a prism
or grating and a scanning mechanism provide better
spectral resolution, but are more complex and, in
turn, more expensive and more subject to misalignment.
    A class of instruments employing gas-cell corre-
lation spectroscopy to provide good sensitivity and
specificity at relatively low cost has been developed
for a variety of applications. Gas-cell correlation
instruments of various types have been discussed in
References 1-5. In one form, the energy beam traversing the sample gas is alternately directed through
either a correlation cell or an attenuator. The
correlation cell, also frequently called a gas filter,
contains the gas species to be measured, and the
attenuator has nearly constant transmittance at all
wavelengths of interest. An interference filter
restricts the spectral bandpass to a region of absorption by the gas species being measured. The gas
in the correlation cell is essentially opaque at the
wavelengths of maximum absorption for the gas species
being measured. Therefore, as the correlation cell
and the attenuator are alternated in and out of the
beam, the energy at wavelengths of strong absorption
by the gas are modulated.
    In the following section, some very simple models
of absorption spectra are used to describe the
spectroscopic principles of detection and discrimination of gas-cell correlation spectrometers. The
illustrations used are helpful in understanding many
of the factors involved in designing a gas-cell
correlation instrument for a particular application.

Brief Description of Instrument

    The primary purpose of the work reported here has
been to perform the required research and to design
and construct a gas-cell correlation instrument for
across-the-stack monitoring of stationary source
emissions. Any of the five gases—NO, CO, SO_2, HF,
or HCl—can be measured over the concentration
ranges ordinarily of interest for stationary sources.
SO_3 was originally included, but subsequent research

indicated that this gas could probably be measured better by methods other than gas-cell correlation spectroscopy.

Conversion of the instrument from one gas species to another is relatively simple and involves changing a correlation cell and a bandpass filter. Changing the radiant energy source and/or the detector may also be required, depending on the gas species involved. The instrument has been designed for versatility; it is necessarily more complex than an instrument designed to use the same principle of operation for a single gas species on a particular type of stack. In some cases, the stability, the signal-to-noise ratio and the ease of operation have been compromised in order to retain the versatility. The instrument is intended primarily as a tool for the Environmental Protection Agency to evaluate gas-cell correlation methods for a variety of across-the-stack measurements. By testing it under various conditions with different instrument parameters, valuable information can be obtained for the design of simpler, single-purpose gas-cell correlation instruments and to determine the factors that limit their performance.

By combining gas-cell correlation techniques along with dispersion techniques to carefully control the spectral bandpass, it has been possible to greatly improve the discrimination and the stability. A small, permanently aligned assembly employing a grating as the dispersing element is included for each of the five gases with the spectral bandpass selected to give optimum discrimination, stability, and signal-to-noise ratio. Each assembly uses a multiple-slit grid in the focal plane of the dispersed radiant energy to pass narrow spectral intervals that contain strong absorption by the gas. Discrimination and stability are improved by rejecting much of the energy at wavelengths where the gas does not absorb strongly. The improved discrimination against H_2O is particularly important when measuring NO. The instrument can be used with a grating assembly or without it, as in the more conventional manner.

Stack diameters between 1.5 m and 10 m can be accounted for with only minor adjustments of a few optical components. In normal operation the stack is "double-passed" by using a retro-reflector on the side opposite the radiant energy source to reflect the energy back through the stack to the remainder of the instrument. However, by moving the source and some associated optics to the opposite side, it is possible to "single-pass" the stack.

At the time of this writing, the only test data for this instrument have been obtained in the laboratory with synthetic samples of effluent. However, a simple demonstration model has been tested on the stack of a coal-burning power plant where transmittance curves were also scanned with a small grating spectrometer in order to measure the concentrations of known pollutants and to check for the presence of others. The data obtained on the stack and in the laboratory indicate that gas-cell correlation instruments can be built with adequate sensitivity and discrimination to monitor any of the five gases listed above. The ultimate performance of any instrument of this type for across-the-stack measurements will probably be limited by the uncertainty of determining and maintaining the "zero-balance" that corresponds to no absorbing gas in the stack. This is a much more serious problem for across-the-stack instruments than for instruments in which the sample area can be evacuated or flushed with a nonabsorbing gas.

Definitions, Symbols and Units

T transmittance at a particular wavenumber (or wavelength) that would be measured with a spectrometer having infinite resolving power. Subscripts indicate the transmittance of different components or gas samples. $\bar{T}$ indicates the average transmittance over a specified interval.

T_{sam} transmittance of the sample gas being analyzed

T_x transmittance of the gas species x to be measured in the sample gas. This is equivalent to T_{sam} if only species x is present in the sample cell.

T_y transmittance of an interfering gas of species y in the sample gas that may produce a false indication of species x

T_f transmittance of the spectral bandpass filter

T_L transmittance of the gas of species x at low pressure in a correlation cell

T_{att} transmittance of an attenuator, assumed to be constant over the spectral bandpass of interest

T_H transmittance of the gas of species x at high pressure in a correlation cell

ν wavenumber of radiant energy (cm^{-1})

ν_o wavenumber of the center of an absorption line. Other subscripts or superscripts indicate particular values (cm^{-1}).

λ wavelength of radiant energy expressed in micrometers (μm)

f_c carrier frequency (high) (Hz)

f_a alternator frequency (low) (Hz)

p partial pressure of a particular gas species, usually species x (atm)

P total pressure of a gas mixture (atm)

ℓ geometrical path length of the radiant energy beam through a gas sample (cm)

C p/P, concentration of a gas species, usually expressed in parts per million (ppm) or per cent

u, w $p\ell$ (atm cm), absorber thickness of a particular gas species. u refers to gas in the sample cell and w to gas species x in a correlation cell. One atm cm is equivalent to 10^4 ppm meters.

k absorption coefficient at wavenumber ν of an absorption line $(\text{atm cm})^{-1}$

s $\int k d\nu$, intensity of an absorption line $(\text{atm}^{-1} \text{ cm}^{-1} \text{ cm}^{-1})$

V_c voltage component of amplified detector signal at carrier frequency f_c

V_a voltage component of amplified detector signal at alternating frequency f_a

V' V_a/V_c, normalized voltage. Normalization is made so that $V' = 1$ when there is 100% modulation of the beam at frequency f_a. See Equation (5) and related text.

D.R.   discrimination ratio, ratio of the concentration of an interfering gas species to the concentration of the species being measured that produces the same reading. This ratio may be positive or negative.

F $\lim\limits_{u\to 0} \dfrac{V'}{u}$, Pembrook Factor $(\text{atm cm})^{-1}$

E chopped radiant energy. Equations (1) and (2) and related text define E_L and E_{att} and explain the subscripts.

M constant relating radiant energy to transmittances. See Equations (1) and (2).

Δ and Σ are defined and explained by Equations (3) and (4) and related text.

SPECTROSCOPIC PRINCIPLES

Correlation in Simple Spectral Models

Gas-cell correlation spectrometers of the type described here depend for their sensitivity on the correlation between the structure in the spectrum of the gas species x to be measured in the sample cell and that of the gas in a correlation cell. The desired correlation is achieved by filling the correlation cell with the same gas species x. Generally, the spectral bandpass of the instrument includes several absorption lines so that the gas in the correlation cell has large fluctuations in the transmittance at different wavenumbers. Figures 10.1-10.4 are based on a simple instrument and a greatly simplified model spectrum to illustrate the spectroscopic principles of detection and discrimination in such a way that the reader can easily follow the mathematics involved.

An optical diagram of a simple gas-cell correlation spectrometer is illustrated in Figure 10.1. Radiant energy from a source passes through either a correlation cell or an attenuator, then through the sample cell and a bandpass filter to the detector. The sample cell may be a stack or a closed cell with windows, or it may simply be an open space such as an atmospheric path. The correlation cell is filled with gas species x to be measured. The attenuator

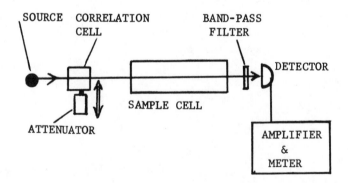

Figure 10.1. Simplified optical diagram of a basic gas-cell correlation spectrometer.

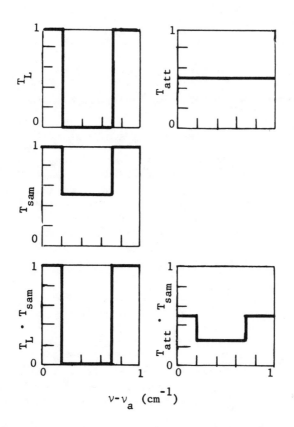

$$\nu - \nu_a \ (cm^{-1})$$

Figure 10.2. *Plots of transmittance for a model gas illustrating the spectroscopic principles of detection. The average transmittance is as follows:*

Upper panels, $\overline{T}_L = 0.50$ *and* $\overline{T}_{att} = 0.50$.

Middle panels, $\overline{T}_{sam} = 0.75$.

Lower panels, $\overline{T_L \cdot T_{sam}} = 0.50$ *and* $\overline{T_{att} \cdot T_{sam}} = 0.375$.

and the correlation cell are alternately moved into the beam at the alternator frequency f_a. An ac amplifier measures the component of the detector signal at frequency f_a that results from a difference between the amounts of radiant energy on the detector during the two halves of the alternator cycle.

For simplicity, we assume that the transmittance of the bandpass filter is unity from wavenumber ν_a

to $\nu_a + 1$ cm^{-1}, and is zero elsewhere. A transmission spectrum of the gas in the correlation cell is illustrated in the upper-left corner of Figure 10.2. The upper-right portion shows the transmittance spectrum of the attenuator over the same spectral region. Note that $\overline{T}_L$, the average transmittance of the correlation cell, is made equal to $\overline{T}_{att}$, the average transmittance of the attenuator. When there is no absorbing gas in the sample cell, the amplifier output is zero because the amount of radiant energy incident on the detector is the same during both halves of the alternator cycle. A sample of gas of species x in the sample cell is represented by the middle panel of Figure 10.2. Because the same gas species exists in both the correlation cell and the sample cell, the absorption occurs over the same spectral interval in both cells. The difference in the transmittance results from different amounts of gas in the two cells.

The transmittance of the correlation cell and the sample in series is given by $T_L \cdot T_{sam}$. In this case the sample does not affect the transmittance because the correlation cell absorbs all of the energy at wavenumbers where the sample absorbs. During the correlation-cell half of the alternator cycle, the energy on the detector is proportional to $\int T_L \cdot T_{sam}$ which is numerically equivalent to $\overline{T_L \cdot T_{sam}}$ in the 1 cm^{-1} interval. The lower right-hand portion of Figure 10.2 illustrates the spectrum of the energy during the attenuator-half of the cycle. Note that the two halves are no longer balanced with the sample in the sample cell.

$$\overline{T_L \cdot T_{sam}} = 0.50 > \overline{T_{att} \cdot T_{sam}} = 0.375$$

The difference is measured by the ac amplifier and meter and is related to the concentration of the gas in the sample. Because T_{att} is a constant equal to $\overline{T}_L$, $\overline{T_L \cdot T_{sam}} = \overline{T_{att} \cdot T_{sam}} = \overline{T}_{att} \cdot \overline{T}_{sam}$. When there is a positive correlation between the spectral structures of the gases in correlation cell L and the sample cell, $\overline{T_L \cdot T_{sam}} > \overline{T}_L \cdot \overline{T}_{sam}$, as in the example above. If $\overline{T_L \cdot T_{sam}} < \overline{T}_L \cdot \overline{T}_{sam}$, the correlation is negative. The two terms are equal when there is no correlation.

Figures 10.3 and 10.4 are based on the same correlation cell and attenuator as the previous example and illustrate the interference by two

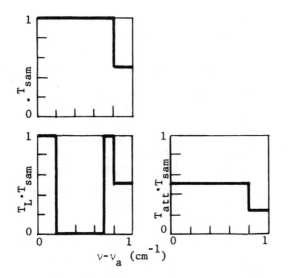

Figure 10.3. *Plots of transmittance illustrating interference by a gas species other than the one being measured.*

Upper panels, $\overline{T}_{sam}$ = 0.9.

Lower panels, $\overline{T_L \cdot T_{sam}}$ = 0.40 and $\overline{T_{att} \cdot T_{sam}}$ = 0.45.

different absorbing gas species other than species x in the sample. The gas species illustrated in the upper panel of Figure 10.3 absorbs, but at different wavenumbers than does species x. The correlation between the spectra of the sample and species x is obviously negative with the resulting misbalance opposite to that when only species x is in the sample. $\overline{T_L \cdot T_{sam}}$ = 0.40 < $\overline{T_{att} \cdot T_{sam}}$ = 0.45. If the interfering gas is alone in the sample cell, as indicated by Figure 10.3, the instrument will indicate a negative amount of species x. When both species x and the interfering gas are present, the indicated concentration of species x is too low.

The interfering gas of species y illustrated in Figure 10.4 has two absorption lines, one inside the absorption line of species x and one outside. No correlation exists between the spectra of this gas and species x, and $\overline{T_L \cdot T_y}$ = 0.4 = $\overline{T_{att} \cdot T_y}$. The interfering gas decreases the energy incident on it by the same fraction during both halves of the cycle, and the system remains balanced. The lower panels

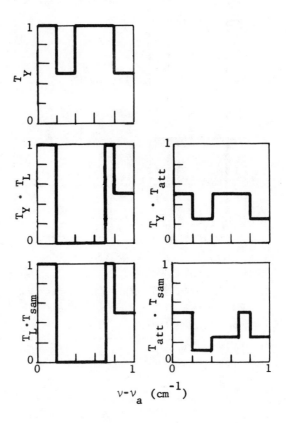

Figure 10.4.   Plots of transmittance illustrating the effect of
an absorbing gas of species y whose spectrum is
not correlated with that of species x.   The upper
panel represents gas species y alone in the sample
cell.   The combined transmittances of species y
with the correlation cell and with the attenuator
are shown in the two middle panels.   The corresponding two plots in the two lower panels are for
a sample consisting of both species x and species y

Upper panel, $\overline{T}_y = 0.80$.

Middle panels, $\overline{T_L \cdot T_y} = 0.40$ and $\overline{T_{att} \cdot T_y} = 0.40$.

Lower panels, $\overline{T_L \cdot T_{sam}} = 0.40$ and $\overline{T_{att} \cdot T_y} = 0.30$

of Figure 10.4 correspond to a sample consisting of species y, whose transmittance is plotted in the upper panel, and species x, represented by the middle panel of Figure 10.2.

We note that $\overline{T_L \cdot T_{sam}} - \overline{T_{att} \cdot T_{sam}} = 0.4 - 0.3 = 0.1$. This difference is less than the 0.125 produced by the same amount of species x, but with no species y present (Figure .2). Although the interfering gas y by itself does not produce a misbalance, it reduces the misbalance produced by species x to 0.8 its original value. This factor is, of course, equal to $\overline{T_y}$.

A similar reduction in the misbalance due to a sample of species x would be produced by continuum absorption or by dirt on the windows of the sample cell. The reduction in the misbalance and in the resulting detector signal can be accounted for by processing the detector signal in such a way that the output is proportional to $(\overline{T_L \cdot T_{sam}} - \overline{T_{att} \cdot T_{sam}})/(\overline{T_L \cdot T_{sam}} + \overline{T_{att} \cdot T_{sam}})$. For example, from Figure 10.2, we see that this ratio for species x, $(0.50 - 0.375)/(0.50 + 0.375) = 1/7$, is equal to the corresponding ratio for the same quantity of species x plus the uncorrelated species y represented by Figure 10.4: $(0.40 - 0.30)/(0.40 + 0.30) = 1/7$.

The instrument described in the following section measures a quantity proportional to this ratio. In most real situations, the spectral band includes several absorption lines of the gas being measured and may include several lines of one or more interfering gases. If the positions and intensities of the lines of the interfering gas are completely uncorrelated with those of species x, the interfering gas will behave as the one illustrated in Figure 10.4 and not produce a false indication. In practice, the spectrum of a gas absorbing within the spectral bandpass will usually have some correlation, either positive or negative, with that of species x and consequently will interfere with the measurement.

Energy and Voltage Relationships

Figure 10.5 illustrates a gas-cell correlation spectrometer that is more complex than the basic one shown in Figure 10.1. It is similar in operating principle to the one built by us for across-the-stack measurements and can be used with two correlation cells or with one correlation cell and an attenuator.

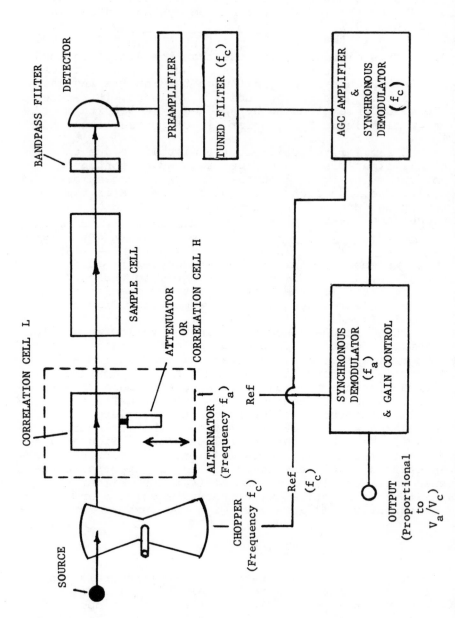

Figure 10.5. Simplified diagram of a gas-cell correlation spectrometer that employs high-frequency

The alternator consists of correlation cell L and either an attenuator or correlation cell H along with the optical and mechanical components that alternately direct the beam through them. The one-cell attenuator system employs correlation cell L and an attenuator. In the two cell system, the attenuator is replaced by correlation cell H, which contains the gas species to be measured at high pressure. In the alternator, the beam may remain fixed while the correlation cells (or attenuator) move as indicated, or the correlation cells may remain fixed while the beam is alternately directed through them by moving mirrors. In either case, the beam passes through the sample cell during both halves of the attenuator cycle.

Before passing through the alternator, the energy beam is chopped at a relatively high frequency f_c, called the carrier frequency. The difference in the energy during the two halves of the alternator cycle modulates the amplitude of the carrier signal. Employing the high frequency chopping in addition to the low frequency modulation makes it possible to measure a quantity proportional to V_a/V_c, where these voltages are proportional to the components of the detector signal at frequencies f_a and f_c, respectively. The ratio V_a/V_c can be related experimentally to the gas concentration and is relatively independent of the source brightness, detector sensitivity, dirt on windows or mirrors, etc.

Figure 10.6 contains several plots of calculated transmittance for a model spectrum used to illustrate the spectroscopic principles of the one-cell attenuator system illustrated in Figure 10.5. The curves correspond closely to those of Figure 10.2, except that the rectangular-shaped spectral lines of Figure 10.2 are replaced by two absorption lines with spacings, intensities, and half-widths typical of a portion of the fundamental CO band. The curves of Figure 10.6 are quite realistic, but the effects of correlation are less obvious than in Figure 10.2. For simplicity, we assume that the bandpass filter passes only the spectral region between the centers of the absorption lines ($T_f = 1$ for $0 \le \nu - \nu_a \le 4$ cm^{-1}, and $T_f = 0$ for all other ν).

The radiant energy chopped at frequency f_c incident on the detector during the half cycle that the correlation cell L is in the beam is given by

$$E_L = M \int T_f T_L T_{sam} \, d\nu \qquad (1)$$

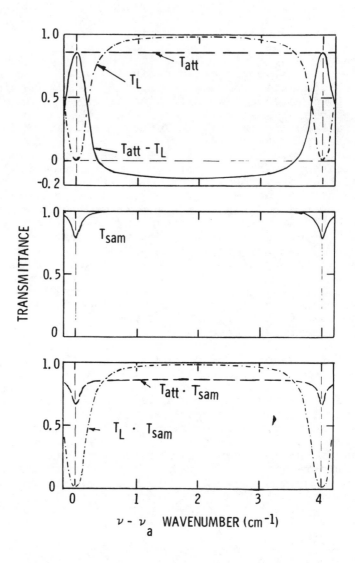

Figure 10.6.   *Calculated spectral plots of transmittance for a model spectrum similar to a portion of a CO band. The curves illustrate the principle of detection of the one-cell attenuator system.   The plot of T_L is based on a correlation cell containing 0.2 atm cm of CO at 1 atm pressure.*

The corresponding energy during the attenuator-half
of the cycle is

$$E_{att} = M \int T_f\, T_{att}\, T_{sam}\, d\nu \qquad (2)$$

Although T_f is constant in the present example, it
is left under the integral sign so that the equations
will be valid for all situations. The constant M
depends on source intensity and size, aperture sizes,
window transmittance, mirror reflectivities, etc.
For simplicity, we assume that the spectral emissivity
of the source and sensitivity of the detector are both
constant over the spectral interval of interest. The
remainder of the discussion on spectroscopic principles is based on the condition that $E_L = E_{att}$ with
no sample in the sample cell ($T_{sam} = 1$).

The curve in the upper panel of Figure 10.6
labeled ($T_{att} - T_L$) represents the spectral response
of the system at frequency f_a. Sample absorption
where ($T_{att} - T_L$) > 0 produces a positive V_a. On
the other hand, a gas in the sample cell with a strong
absorption line where ($T_{att} - T_L$) < 0 and no absorption where this quantity is > 0 would have a negative
correlation with the gas in the correlation cell and
would produce a negative V_a. If the spectrum of the
sample gas is completely uncorrelated with the
spectrum of the gas in the correlation cell, the
absorption producing negative V_a balances that
producing positive V_a.

The spectrum shown in the center panel of Figure
10.6 corresponds to 0.01 atm cm of CO at 1 atm pressure in the sample cell. During the half of the
cycle when correlation cell L is in the beam, the
transmittance of the beam that has passed through
the sample is given by $T_L \cdot T_{sam}$. This quantity is
indicated in the lower panel along with the corresponding curve that applies during the attenuator-half
of the alternator cycle. The sample absorbs only a
very small portion of the radiant energy that has
passed through correlation cell L because little
energy remains at the wavenumbers where the sample
absorbs. During the attenuator-half of the cycle,
the sample absorbs a larger portion of the energy
with the result that $E_L > E_{att}$. A detector voltage
component V_a proportional to ($E_L - E_{att}$) appears as
a modulation of the carrier voltage V_c.

The curves in Figure 10.7 illustrate the dependence of E_{att} and E_L on sample absorber thickness

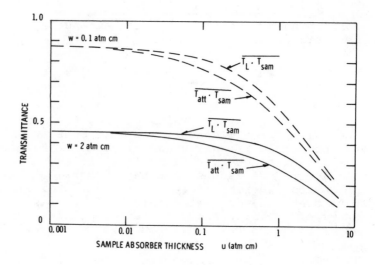

Figure 10.7. *Plots of calculated* $\overline{T_L \cdot T_{sam}}$ *and* $\overline{T_{att} \cdot T_{sam}}$ *vs. sample absorber thickness. The curves are based on the spectral model similar to CO that is illustrated in Figure 10.6.*

for the spectral model illustrated in Figure 10.6. Each pair of curves corresponds to a different absorber thickness w of species x in the correlation cell; the pressure is maintained at 1 atm. Because $T_f = 1$ over the region of interest, $0 \leq \nu - \nu_a \leq 4$ cm^{-1}, $\overline{T_L \cdot T_{sam}} = E_L/4M$, and $\overline{T_{att} \cdot T_{sam}} = E_{att}/4M$ As the sample absorber thickness u approaches zero, T_{sam} approaches unity, and $\overline{T_L \cdot T_{sam}}$ and $\overline{T_{att} \cdot T_{sam}}$ approach $\overline{T_L} = \overline{T_{att}}$.

We define two useful quantities:

$$\Delta = \int T_f (T_L - T_{att}) T_{sam} \, d\nu = (E_L - E_{att})/M \qquad (3)$$

$$\Sigma = \int T_f (T_L + T_{att}) T_{sam} \, d\nu = (E_L + E_{att})/M \qquad (4)$$

For the spectral model of Figure 10.6, Δ is proportional to V_a and to the difference between values plotted in Figure 10.7 for given values of u and w. Similarly, Σ is proportional to the sum of the same values and to the average value of V_a. Neither Δ

nor Σ is easy to measure directly; however, the ratio
can be measured with an instrument of the kind illustrated in Figure 10.5. In order to calibrate the
instrument, the radiant energy beam through the
attenuator is temporarily blocked so that T_{att} and,
in turn, E_{att} equal zero. Under this condition
$\Delta/\Sigma = 1$, and the amplifier gain settings are adjusted
so that the output voltages V_a and V_c are equal. We
define

$$V' = V_a/V_c \qquad (5)$$

with the gain settings adjusted as just described.
If the gain settings remain fixed, it follows that
$V' = V_a/V_c = \Delta/\Sigma$ when the attenuator beam is unblocked
and the voltages are measured.

An automatic gain control circuit that maintains
V_c constant can further simplify the measurement.
When this is used, V' is proportional to V_a, which
is measured directly. Thus, only one voltage, V_a,
need be read in order to determine Δ/Σ.

Curves based on calculated values of $V' = \Delta/\Sigma$
are shown in Figure 10.8 for five values of w, including the two represented in Figure 10.7. Recall

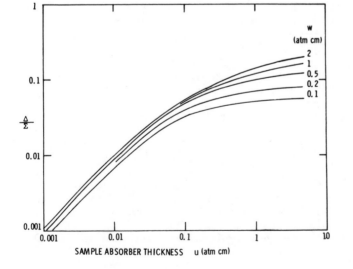

*Figure 10.8. Plot of calculated Δ/Σ vs. sample absorber thick-
ness. Each curve is identified by w, the absorber
thickness of gas species x in correlation cell L.
$\overline{T}_L$ for each value of w is: 2 atm cm, 0.461; 1 atm
cm, 0.600; 0.5 atm cm, 0.711; 0.2 atm cm, 0.817;
and 0.1 atm cm, 0.875.*

that the calculated curves are based on a portion of
a model spectrum similar to a CO spectrum and that
the spectral interval corresponds to the spacing
between two adjacent lines. Because the curves of
Figure 10.8 are based on ratios of integrated transmittance, it follows that the same curves would be
obtained if the spectral interval of the model
spectrum were widened to include several identical
lines with the same spacings. Such a model is
similar to the eight or ten strongest lines in
either the P branch or the R branch of the fundamental CO band. We have employed a gas-cell correlation spectrometer with a spectral bandpass that
included approximately ten of the strongest lines
in the R branch of the CO band. Each experimental
curve we obtained was very similar to the one for
the corresponding value of w in Figure 10.8. The
differences between the experimental and the calculated curves could be explained by the difference
between the model spectrum and the real CO spectrum
and by the shape of the transmission curve of the
bandpass filter used in the instrument. A "rectangul
bandpass in which T_f = 0 or 1 has been assumed for
the calculations.

An experimental curve of V_a *vs.* u can be obtaine
without adjusting the amplifier gains by the procedur
described above. The shape of the curve on a logarit
mic scale is the same regardless of changes in the
gain setting, which shift the curve vertically.
However, adjusting the gains as outlined makes it
possible to relate V_a directly to Δ/Σ. For small
values of u, the curves of Figure 10.8 have unity
slope, indicating that Δ/Σ is proportional to u.
Furthermore, for a given value of u, Δ/Σ varies by
a factor of only about 1.5 for variations of a factor
of 20 for w.

If w were further reduced to values much less
than those indicated, T_L would no longer approximate
zero near the line centers, and Δ/Σ would reduce
more rapidly. In many situations, the major factor
limiting the accuracy of a gas-cell correlation instrument is "zero-drift," the slow variation in the
signal output when $u = 0$. An important instrumental
parameter is given by

$$F = \lim_{u \to 0} \frac{V'}{u} = \lim_{u \to 0} \frac{\Delta}{u\,\Sigma} \quad (\text{atm cm})^{-1} \qquad (6)$$

For lack of an unambiguous and descriptive name, we have named F the Pembrook Factor after one of our colleagues, John Pembrook, who has done considerable work in developing methods to evaluate instrument performance.

From Figure 10..8 we see, for example, that $\Delta/\Sigma \simeq 0.001$ when u = 0.001 atm cm and w = 1 atm cm. Thus, F = 1.00 (atm cm)$^{-1}$, and a zero-drift corresponding to a misbalance of Δ/Σ = 0.001 produces an error equivalent to 0.001 atm cm of the gas being measured. From Equations (3) and (4), we see that this corresponds to a zero-drift such that 0.998 as much chopped radiant energy is transmitted through the alternator during one-half of its cycle as during the other half [(1 - 0.998)/(1 + 0.998) $\simeq$ 0.001]. It follows that the spectral bandpass and correlation cell parameters should be chosen to produce a large value of F if the sample cannot be removed so that zero drift can be checked and accounted for regularly.

The slopes of the curves of Figure 10.8 decrease with increasing u. For values of u more than one or two times w, the dependence of V' on u is so slight that it establishes a practical upper limit of u for an instrument.

Design Factors

From Figures 10.6, 10.7, and 10.8 we can establish some general "rules of thumb" and factors to be considered in determining the optimum spectral bandpass, sample cell length, and correlation cell parameters for a particular application. The spectral bandpass should be chosen to include strong absorption lines of gas species x to be measured and a minimum of absorption by other gas species that may be present in the sample gas. Interference by other gas species generally can be decreased by employing a selected narrow bandpass. However, a lower practical limit on the width of the spectral bandpass may be determined by filters that are available or by detector noise. If the bandpass is very narrow, the signal output, which is proportional to Δ, [Equation (3)] for small u may be less than the detector noise.

The absorber thickness w in the correlation cell should be great enough that the gas is essentially opaque at the wavenumbers of strong absorption near the line centers. The curves of Figure 10.8 show that by employing a relatively large w, a larger Pembrook Factor F can be achieved along with an extension of the dynamic range to larger values of u.

Two factors establish a practical upper limit on w.
If the correlation cell is opaque over too wide a
spectral interval near each line center, the "sharp"
structure is lost and discrimination against other
gases may suffer. Furthermore, as w increases to
the point that $\overline{T}_L$ is small, the signal output resulting from a small u decreases, and detector noise
may become a problem. The decrease in signal output
with increasing w can be observed in Figure 10.7, in
which Δ is proportional to the separation between a
pair of curves. For a given u less than approximate
0.1 atm cm, Δ is larger for w = 0.1 atm cm than for
w = 2 atm cm. If w were increased further, Δ would
continue to decrease, but the ratio Δ/Σ would continue to increase slightly as indicated in Figure
10.8 because Σ would also decrease.

The optimum combination of sensitivity and discrimination are generally achieved when the pressure
of the gas in the low pressure correlation cell is
equal to or somewhat less than the pressure in the
sample cell. This ensures that the widths of the
pressure-broadened lines are approximately the same
in both cells. We have shown experimentally and
analytically that the performance is seldom affected
significantly when the correlation cell pressure
varies from the sample pressure down to less than
half of its value. If the spectral bandpass contains
only spectral lines of approximately equal intensity
the dependence on pressure is less than if the lines
cover a wide range of intensities. It is understood
that as the pressure is varied, the absorber thickness
is adjusted to maintain near opacity near the centers
of the strong lines. The desired combination of total
pressure and absorber thickness is achieved by varying
the length of the correlation cell or by adding a
nonabsorbing gas such as N_2 to collision broaden the
lines.

For a fixed sample cell length, the range of
concentrations that can be measured is limited on
the low end by noise or by uncertainty in the correct
zero reading and on the upper end by the "saturation"
indicated by the curves of Figure 10.8. If it is
practical, the sample cell length is selected to be
consistent with the range of concentrations to be
measured. Two or more sample cell lengths may be
required to cover the entire range of concentrations.
In some applications, such as across-the-stack
measurements of stationary source pollutants, the
optical length through the sample gas may be difficult
to change. Double-passing the energy beam through a

stack to increase the length may be practical, but limits on window sizes and on the mechanical stability of the optical platforms make it difficult to pass the beam across the stack more than twice. If one pass of the stack produces too much absorption, it may be necessary to select a different spectral bandpass in which the absorption is weaker. Another solution is to use an optical arrangement that passes the radiant energy beam across only part of the stack. This method does not provide the averaging of a path that extends across the entire stack.

Narrowing the absorption lines by decreasing the sample pressure to less than 1 atm can greatly improve the discrimination and sensitivity for some gases. This is particularly true for SO_2, whose absorption lines are so closely spaced that they overlap and smooth out much of the spectral structure when the sample is near 1 atm. At pressures less than about 0.1 atm, the spectral structure of SO_2 is greatly increased so that the performance of a gas cell correlation spectrometer is improved. The optimum sample cell length and correlation cell parameters for any application obviously depend on the sample pressure. The half-width of a collision-broadened absorption line is proportional to the total gas pressure and inversely proportional to the square root of the temperature. Therefore, the lines of a hot gas in a smokestack are narrowed by the higher temperature although the pressure is approximately atmospheric.

Two-Cell System

Two two-cell system is illustrated schematically by Figure 10.5 when a high-pressure correlation cell H is used in place of the neutral density attenuator. Figure 10.9 illustrates the spectroscopic principles of this system with the same low-pressure sample in correlation cell L as in the case illustrated by Figure 10.6. The model is based on correlation cell H filled to 5 atm with CO, the gas to be detected. The absorber thickness of the CO in this cell, 0.108 atm cm, was adjusted so that $\int T_f T_H d\nu = \int T_f T_L d\nu$. Because of the pressure broadening of the absorption lines, the gas in cell H absorbs less near the centers of the lines, but more in the wings of the lines, than does the low-pressure gas in cell L. The curve labeled $(T_H - T_L)$ represents the spectral response of the system at frequency f_a. By comparing Figure 10.9

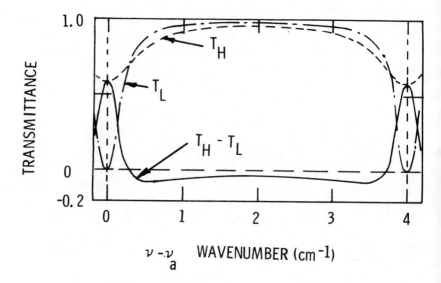

Figure 10.9.   *Calculated spectral plots of transmittance for a model spectrum illustrating the principle of detection of the two-cell system.*

with Figure 10.6, we see that the response $|T_H-T_L|$, of the two-cell system is less than $|T_{att}-T_L|$ of the one-cell attenuator system in the region midway between the lines. Thus, the two-cell system would discriminate better against an absorption line of an interfering gas in this region. The two-cell system is only about three-fifths as sensitive as the one-cell attenuator system because of the smaller value of (T_H-T_L) near the centers of the lines.

In matching $\int T_f T_L d\nu$ with $\int T_f T_H d\nu$, the pressure of the gas in cell H could have been made higher with a corresponding shorter path to maintain the balance between the two beams. This change would further broaden the lines, increasing T_H near the line centers and decreasing it in the wings. As a result, the spectral response would be greater near the centers of the lines. However, this would be accompanied by an increased negative response in the wings of the lines, causing the system to respond more to absorption by other gases with absorption lines between the CO lines. The optimum pressure in the cells depends on the spectra of the gas species to be measured and any interfering gases in the sample. In many cases, the decrease in response of the two-cell system is more than compensated for by an increased discrimination over the one-cell attenuator system.

Figure 10.10 illustrates model spectra for which the discrimination by the two-cell system is far superior to that by the one-cell attenuator system.

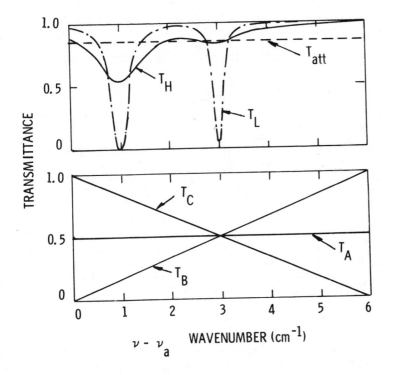

Figure 10.10. *Calculated spectral plots of transmittance demonstrating the principles of discrimination against continuum absorption by the two-cell and the one-cell attenuator systems.*

The curves labeled T_L and T_H in the upper panel correspond to the transmittances of low and high pressure correlation cells, respectively. Both have the same average transmittance, which is represented by the curve labeled T_{att}. Thus the two curves T_L and T_H correspond to the two-cell system, and the curves T_L and T_{att} correspond to the one-cell

attenuator system. Both of the absorption lines
have a half-width α of 0.05 cm^{-1} for the low pressure
sample, and 0.5 cm^{-1} for the high pressure sample.
The intensities of the lines at $(\nu-\nu_a) = 1$, and
3 cm^{-1} are 4 atm^{-1}cm^{-1}cm^{-1} and 1 atm^{-1}cm^{-1}cm^{-1},
respectively. The absorber thicknesses of gas x in
the low and high pressure cells are 0.5 and 0.24074
atm cm, respectively. It is assumed that $T_f = 1$
for $0 \leq (\nu-\nu_a) \leq 6$ cm^{-1} and $T_f = 0$ for all other ν.
      The three curves in the lower panel represent
three simple types of continuum absorption by interfering gases in the sample. Table 10.1 summarizes
the relative sensitivities of the two systems to the
three types of continuum and to a small amount of
gas species x at the same pressure as the gas in
correlation cell L. The equivalent error corresponds
to the amount of gas species x that produces the same
Δ/Σ as that produced by the interfering continuum.
Continuum A has constant transmittance and produces
no error, whereas continua B and C produce signals
in the one-cell attenuator system that correspond
to a sample containing approximately 0.108 atm cm
of the gas x. The equivalent error for the two-cell
system is only about 2% as much. Sample B produces
an error that corresponds to a positive amount of
gas x because the general slope of the transmittance
curve is similar to that of gas x; *i.e.*, the transmittance increases with increasing wavenumber for
both gas x and continuum B. As expected, the opposite
is true for continuum C, whose transmittance curve
slopes in the opposite direction.

Table 10.1

*Comparison of the Sensitivity and Discrimination
of Two-Cell and One-Cell Attenuator Systems*

Sample	Two-Cell		One-Cell Attenuator	
	$\dfrac{\Delta}{\Sigma}$	Equivalent Error (atm cm)	$\dfrac{\Delta}{\Sigma}$	Equivalent Error (atm cm)
0.001 atm cm of gas	0.0002066		0.0003203	
Continuum A	0	0	0	0
Continuum B	0.0004778	+0.00231	0.03474	+0.1084
Continuum C	-0.0004778	-0.00231	-0.03474	-0.1084

A small amount of gas species x produces a smaller signal in the two-cell system than in the one-cell attenuator system; however, the increase in discrimination may be more important than the loss in sensitivity. Similar results would be obtained if the continua were replaced by spectra with the same general slopes, but consisting of many lines. Thus, curves B or C may be representative of the spectra near the edge of an absorption band of an interfering gas. The model spectrum of gas x in the correlation cells is also typical of the edge of many absorption bands in which the lines get progressively weaker with increasing distance from the band center. As an example, interference by the wing of a CO_2 band when measuring CO is similar to the model illustrated in Figure 10.10. In such a case the discrimination by the two-cell system can be much better than that by the one-cell attenuator system.

Multiple Correlation Cell Systems

The discussion above has been restricted to relatively simple instruments for which the processing of the detector signal is not excessively complicated. By using more than two correlation cells with different amounts of gas at either different pressures or the same pressure, it may be possible to obtain more information about the sample absorption and to account better for interfering gases, particulate matter, and variations in sample temperature or pressure. In general a measurement with a correlation cell containing a small amount of gas at low pressure provides information about the absorption near the centers of the lines. Similarly, a correlation cell with more gas absorbs over a wide interval near the line centers and provides information about absorption in the wings of the lines. An alternate method is to use a single correlation cell and vary the gas pressure in it. Processing the detector signal for an instrument with more than two correlation cells, or with one of varying pressure, is necessarily more complicated than for two cells or for one cell and an attenuator. The added complication is usually not required for absorption measurements, but it may be justified for passive measurements in which the hot gas being measured serves as the radiant energy source.

INSTRUMENT DESIGN AND PERFORMANCE

Optical Layout of Basic Instrument

Figure 10.11 shows an optical diagram of one version of a versatile instrument we have built recently. It employs the principles described above and can be used to compare different methods of gas correlation cell spectroscopy under a variety of situations. Different gas species can be measured by interchanging the gas correlation cells, a bandpass

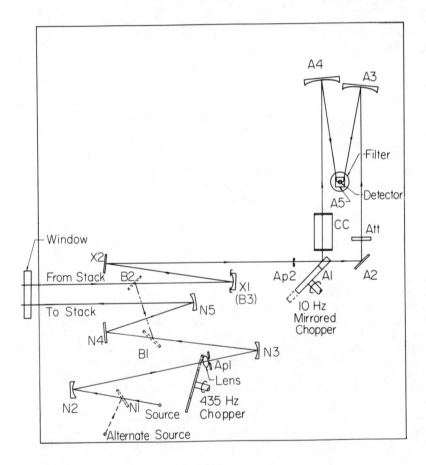

Figure 10.11. Optical diagram of a gas-cell correlation instrument for across-the-stack measurements.

filter, and possibly the source and detector. The
instrument is equipped with correlation cells and
filters to measure NO, CO, SO_2, HCl and HF. All of
the optical components shown in Figure 10.11 are
mounted on a baseplate approximately 47 cm x 53 cm.
Except for a preamplifier, all of the electronic
components are mounted separate from the optics.

In Figure 10.11, mirrors indicated by an *N* are
in the entrance section that directs the beam of
radiant energy into the stack. A tungsten filament
bulb in a quartz envelope serves as the radiant
energy source for wavelengths less than approximately
3.6 μm. Either a Nernst glower or an electrically
heated nichrome wire covered with a ceramic material
serves as a source for longer wavelengths. Two
sources can be kept in place. Mirror N1 is removed
or put in place as shown, depending on the source
being used. In normal operation bypass mirrors B1,
B2, and B3 are not in place. Mirror N2 forms an
enlarged image of the source that overfills aperture
Ap1; a further enlarged image of the source is formed
on mirror N5, which directs the beam through a CaF_2
window into the stack. A field lens near aperture
Ap1 images mirror N2 on mirror N3. A retro-reflector,
described in the following paragraph, placed on the
opposite side of the stack directs the beam back
through the stack and forms another image of the
source on mirror X1. Mirror X1 images mirror C3 of
the retro-reflector onto aperture Ap2 just ahead of
the alternator. A rotating mirrored chopper, A1,
with two blades and two open sectors alternately
directs the beam through one of two paths. One
alternate path contains correlation cell CC, and
the other contains an attenuator. As discussed previously, the attenuator may be replaced by a high-
pressure correlation cell. Mirror A5 directs the
beam upward through a bandpass filter to the detector.
Either of two liquid-nitrogen cooled detectors is
used: PbS for $\lambda \leq 4$ μm, and PbSe for λ between 4
and 6 μm. The bandpass filter passes a spectral band
that contains absorption by the gas being measured.

Figure 10.12 shows an optical diagram of the
retro-reflector assembly, which is a modified version
of a "cat's eye" reflector. Mirror C1 forms an image
of mirror N5 on the small spherical mirror C2, which
is adjusted to correspond to the distance between C1
and N5. Mirror C2 images C1 near C3. Mirrors C1
and C3 are cut from a single spherical mirror and
mounted so that their centers of curvature are at
the same height but separated slightly in an azimuthal

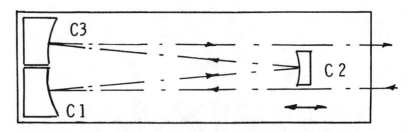

Figure 10.12.   Optical diagram of a retro-reflector assembly
used to redirect the radiant energy beam back
across the stack to the gas-cell correlation
instrument.

direction in order to image N5 on X1. Provided the
relative positions of the three mirrors Cl, C2, and
C3 are rigidly fixed, the position of the image
formed near mirror X1 relative to the effective
source near N5 is insensitive to small angular or
linear displacements of the retro-reflector assembly.
     The radii of curvature of the mirrors are chosen
to be optimum for a stack approximately 3 m in
diameter. For this size of stack, mirrors N2, N3,
Cl, and C3, aperture Ap2, and the detector are conjugate to each other. The source is conjugate to
Apl, N5, C2 and X1.
     The instrument is very sensitive to differences
in the energy passing through the two paths of the
alternator section. Therefore, after the alternator
has been balanced with no absorbing gas in the beam,
it is very important that any changes in aperture or
angular field of view affect both paths of the
alternator in the same way. This is best accomplished
by ensuring that the limits of the beam are well-
defined outside of the alternator and that the
alternator transmits to the detector all of the beam
entering it. The image of C3 at Ap2 is larger than
the aperture,and the image of Ap2 at the detector is
smaller than the sensitive element. Therefore, Ap2
forms the effective aperture as long as any misalignment of the optical components preceding it is
small enough that Ap2 remains completely filled.
The windows and mirrors in the alternator section
are oversized so that a mask in front of mirror X1
limits the angular field-of-view of the detector.
The optical components preceding mirror X1 are designed so that the opening in the mask on mirror X1
is completely filled.

In order to balance the alternator, the bypass mirrors Bl, B2, and B3 are easily positioned as indicated in Figure 10.11 to bypass the stack and provide an optical path with no absorbing gas. Bypass mirror B3 replaces Xl, which has a different radius of curvature and sits in the same position; the same mask is used for Xl and B3. When the bypass optics are in place, Ap2 is conjugate to mirrors N2 and N3, as it is when the beam is traversing the stack. Similarly, B3 is conjugate to the source and Apl as is Xl when it is in place. Thus, the instrument has the same field-of-view while it is being balanced as it has during a measurement. This method of bypassing the stack in order to balance the alternator matches the aperture and field-of-view about as well as is possible without passing the beam through the smoke-stack. Errors due to nonuniformities in the trans-mittance of stack windows or in the reflectivity of mirrors in the entrance section are minimized, but they cannot be eliminated completely. The best balance could probably be achieved by placing a pipe across the stack with the radiant energy beam passing through the pipe. The absorption could be eliminated by flushing the pipe with a nonabsorbing gas. However, this method of balancing is frequently not practical because of space limitations or because of possible interference with the operation of the plant being monitored. It seems probable that the performance of gas-cell correlation instruments for stack measurements ultimately will be limited by the uncertainty in the balance corresponding to no absorption in the stack.

The radiant energy from the source incident upon the detector is chopped at 435 Hz with a 10 Hz modulation that is related to the concentration of the gas being measured. The detector signal is amplified and processed in such a way that the 435 Hz signal serves as a carrier for the 10 Hz signal as indicated by Figure 10.5. In this way, the system is insensitive to 10 Hz variations in the radiant energy on the detector unless the energy is first chopped at 435 Hz. Thus, emission by hot gases in the stack, or by any optical components that follow the 435 Hz chopper, does not influence the instrument output.

Instrument Employing Grating Assembly

The stability and discrimination of an instrument of the type shown in Figure 10.11 could frequently

be improved if the spectral bandpass could be controlled more accurately than can be done with an
interference filter. Figure 10.13 shows another
version of the instrument described above with a
grating assembly to provide the desired bandpass.

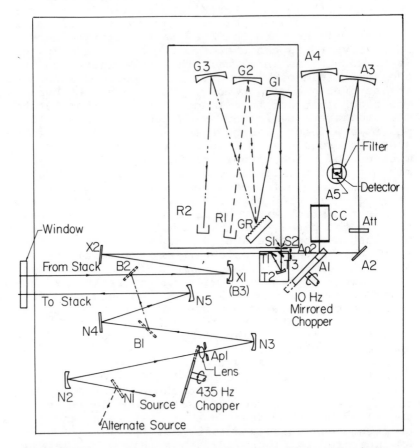

*Figure 10.13.   Optical diagram of a gas-cell correlation
instrument incorporating a grating assembly.*

Either CO or NO can be studied with the grating
assembly shown by blocking the beam at the appropriate
retro-reflector R1 or R2. The bandpass filter near
the detector eliminates overlapping orders of shorter
wavelengths that are passed by the grating assembly.
It is also necessary to change the correlation cell,
and possibly the attenuator, when changing from one
gas to another. A second grating assembly is built

for SO$_2$ or HCl, and a third one serves for HF. Each
grating assembly is constructed in its own box,
which can be interchanged quickly with little or no
realignment of optical components required. The box
for each assembly is mounted to the main base plate
on a three-point system that will not transfer
stresses from the baseplate to the box. All of the
parts in the grating assembly are bonded in place
with epoxy cement to avoid misalignment.

Mirrors T1, T2 and T3 are bonded to a block
that can be removed or put back in place easily.
This transfer optics assembly directs the beam into
an entrance slit S1 that is below the optic axis of
the alternator. The beams of the selected bandpass
exit from the grating assembly through slit S2 at
the height of the optics axis of the alternator.
Mirror T3 directs the beam into the alternator.
When the transfer optics are installed, it is necessary to tilt mirror X1 to direct the beam to the
lower level of mirrors T1, T2 and slit S1. Otherwise,
no adjustments are required when the transfer optics
assembly is installed or removed.

Mirror T2 images mirror X1, and thus the source,
on the entrance slit S1. For a stack diameter of 3 m,
mirror C3 of the retro-reflector is imaged on mirror
T2. As the diameter changes with the resulting change
in distance between mirrors C2 and X1, the conjugate
to mirror C3 moves slightly from mirror T2. A mask
on the grating limits the field-of-view of the
detector.

The grating assembly shown in Figure 10.13 disperses the radiant energy and directs that of wavelengths near 5.25 μm toward the retro-reflector R1.
Energy from a portion of the spectral band selected
for the measurement of NO is redirected back to the
grating, which undisperses it and directs it to
mirror G1 and the exit slit S2. Similarly, wavelengths near 4.6 μm fall on retro-reflector R2, which
returns a portion selected to measure CO. Only one
gas is measured at a time; the retro-reflector not
being used is covered with a mask.

The method by which the radiation is dispersed
and a selected portion is recovered and imaged is
described in Reference 6. The original paper described
the method that uses a prism as a dispersing element;
however, the obvious application to a grating and to
systems with a "tailored" bandpass were pointed out.
Decker[7] has also described an application of the
method to Hadamard spectroscopy.

A view of a retro-reflector and grid assembly
like those indicated by R1 or R2 in Figure 10.13 is

shown in Figure 10.14.   A converging beam of dispersed
radiant energy from mirror G2 (Figure 10.13) is incident upon mirror G4, whose reflecting plane is 45°
from the vertical.   The horizontal grid is in the
focal plane where monochromatic energy forms an image
of the entrance slit S1.   Openings in the grid, such
as the two shown in Figure 10.14, pass the selected
spectral intervals to mirror G5.   Mirrors G4 and G5

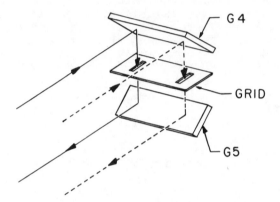

*Figure 10.14.   Optical diagram of a retro-reflector and grid
                assembly.*

are normal to each other so that each ray of the beam
returns to mirror G2 parallel to, but displaced vertically downward from, the ray before it strikes
mirror G4.   The returning beam is completely undispersed by the grating, and the image formed at the
exit slit S2 is undispersed and contains only those
wavelengths passed by the grid in the retro-reflector
and grid assembly.   Because an image of slit S1 is
formed at slit S2, only one of these slits is required
to provide the desired spectral resolution.   Slit S1
is intentionally made somewhat wider than slit S2;
the effective part of S1 is only that portion conjugate to the narrower S2.   Slit S1 reduces scattered
energy reaching the detector; having it oversize
does not degrade the resolution and makes the
positioning of the slits much less critical.

Spectral Properties of the Gases

Figure 10.15 shows two experimental curves of transmittance in the region of the Q-branch and the R-branch of the fundamental NO band. The absorption represented by the upper curve is due to approximately 1.5% H_2O in 8 m of room air. The lower curve, which is displaced downward for clarity, represents the same amount of H_2O and a sample of 0.03 atm of NO in a 10-cm sample cell. The need to reduce inter- ference by H_2O is apparent. Absorption by H_2O is even stronger in the lower-wavenumber P-branch of the NO band. The cross-hatch areas correspond to the six narrow spectral intervals passed by the retro-reflector and grid assembly for NO. Inter- ference by H_2O is reduced by choosing the intervals with a minimum of H_2O absorption. Each interval contains a pair of NO lines.

Blocking the spectral intervals where there is little or no NO absorption decreases Σ with essentially no decrease in Δ [see Equations (3) and (4)]. Thus, the Pembrook Factor F is increased with the resulting increase in the stability of the zero balance. The blocked energy, if allowed to pass, would provide no information about NO concentration, but it would add to any misbalance of the alternator. Information on the positions, intensities, and half-widths of the NO lines can be obtained from two articles by Abels and Shaw.[8,9]

Two transmittance curves of the SO_2 band employed are shown in Figure 10.16. The structure in curve A is due to "clusters" of many closely-spaced lines that are not resolved. When the SO_2 sample is greater than about 1 atm, the lines are broadened enough that they overlap considerably and much of the structure that occurs at low pressures is smoothed out. Con- sequently, the structure in a spectrum of a 1 atm sample that could be observed with a very narrow spectral slitwidth is not greatly different from curve A. Further smoothing of the spectrum with increased pressure is illustrated by curve B of Figure 10.16. At the high pressure, the widths of the SO_2 lines are comparable to the spacings between the clusters of lines.

The relatively "shallow" structure in the spec- trum of SO_2 near 1 atm contributes little to the sensitivity of a gas cell correlation instrument for this gas when the sample is near 1 atm, as in a smokestack. Most of the sensitivity results from the coarse, deep structure rather than from the

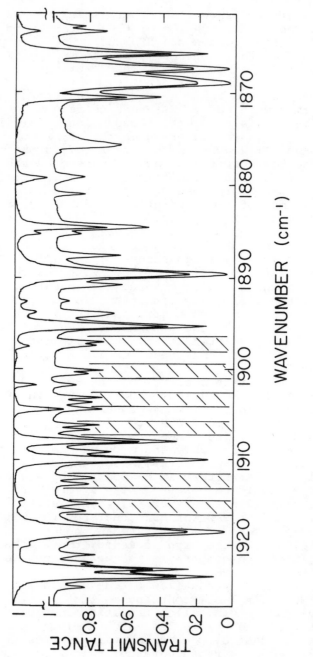

Figure 10.15.   Spectral curves of transmittance of H_2O and of H_2O + NO between 1862 and 1928 cm^{-1}. Spectral slitwidth = 0.35 cm^{-1}.   The transmittance scales of the two curves differ by 0.20.

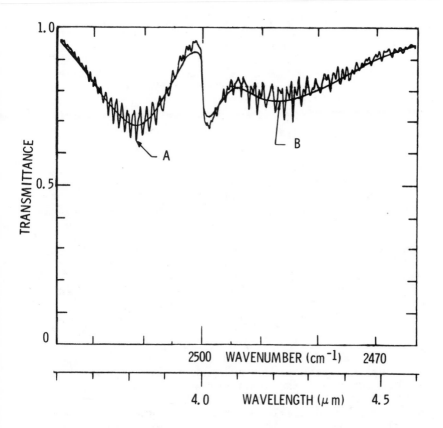

Figure 10.16. *Spectral curves of transmittance of SO_2 between 2465 and 2525 cm^{-1}. The total pressures of the two dilute, room-temperature mixtures of SO_2 in N_2 are 1.00 atm for sample A and 15.0 atm for sample B. u = 0.78 atm cm for sample A; u for sample B was reduced a few per cent to produce nearly the same average transmittance as sample A. The spectral slitwidth ≈ 0.35 cm^{-1}.*

structure due to the line clusters approximately 3 cm^{-1} apart. Because of this, only two slots are employed in the grid for SO_2; one includes a region of weak absorption near 2538 cm^{-1}, and the other includes many lines in the region of strong absorption near 2512 cm^{-1}.

Two bands of SO_2 centered near 1150 cm^{-1} and 1360 cm^{-1} are stronger and would provide higher sensitivity than the one used. However, H_2O interference is much more troublesome and detectors are

less sensitive in the region of the two lower-
wavenumber bands. Data on the absorption by these
bands are given in Reference 10.

Data on the absorption properties of CO, HCl,
and HF are available in a number of reports and
papers. Line parameters used in the design of the
present instrument are from Benedict *et al.*[11] for
CO, Benedict *et al.*[12] for HCl, Meredith and Kent[13]
and Herget *et al.*[14] for HF.

Instrument Performance

Table 10.2 summarizes the important parameters
and the performance of the instrument for the five
gases of interest when the appropriate grating
assembly is used. The second column lists the
spectral intervals passed by the grid in the retro-
reflector and grid assembly. As indicated in the
discussion of Equation (6), the Pembrook Factor given
in the third column relates the absorber thickness
of the gas in the sample to the resulting percentage
misbalance in the two alternator beams.

A Nernst glower source heated by 0.8 amp of
alternating current was used in measuring the absorber
thickness required to produce a signal equivalent to
the peak-to-peak noise arising from the detector.
A three-second time constant was employed. The
laboratory optical path simulated a double-passed
stack 3 m in diameter with 5-cm diameter windows.
Both the PbSe and PbS detectors have 2 mm x 2 mm
sensitive elements cooled with liquid nitrogen.

Water vapor produces the most serious interference when measuring NO. The discrimination ratio
(10,000:1) is based on a 3-m diameter stack containing 5% H_2O. Because of a nonlinear relationship
between the H_2O absorber thickness and the instrument
response, the discrimination ratio depends on the
amount of H_2O. Unrealistically high discrimination
ratios can be observed with high concentrations of
the interfering gas. A meaningful discrimination
ratio should be measured with approximately the same
absorber thickness of the interfering gas as will be
encountered during a measurement.

Several measurements of H_2O interference were
made while measuring NO with an instrument employing
different bandpasses that included several of the
H_2O lines seen in Figure 10.15. Typically, the H_2O
interference was 10-100 times as great as that observed with the grating assembly with a bandpass of
six narrow spectral intervals. By employing the
grating assemblies to carefully select the bandpasses,

Table 10.2

Instrument Parameters and Performance

Gas to be Measured	Spectral Band Pass	Pembrook Factor $(atm\ cm)^{-1}$	Detector and Noise-Equivalent Absorber Thickness $(atm\ cm)$		Remarks on Interfering Gases
NO	6 intervals, each centered on a pair of NO lines near 1900 cm^{-1}	0.73	PbSe 0.003	H_2O	D.R. = 30,000:1 with H_2O at 125°C. 5% H_2O produces error corresponding to 1.6 ppm NO. The 6 pairs of NO lines in the spectral bandpass are relatively free of H_2O absorption. Inclusion of other lines would increase the interference by H_2O.
CO	11 intervals, each centered on a CO line between 2150 and 2200 cm^{-1}	1.75	PbSe 0.0008	CO_2	Interference too small to measure. D.R. > 200,000:1 with CO_2 at 125°C. 10% CO_2 produces error corresponding to less than 0.5 ppm CO.
SO_2	2 intervals, near 2538 and 2512 cm^{-1}	0.088	PbS 0.002	H_2O	D.R. ≃ ~60,000:1 with H_2O at 125°C. 5% H_2O produces error corresponding to ~0.8 ppm CO. No significant interference by other gases in stacks of coal-burning plants in this spectral region.
HCl	3 intervals, each includes corresponding lines of ^{35}Cl and ^{38}Cl isotopes. Intervals are between 2750 and 2850 cm^{-1}	0.58	PbS 0.0005		Intervals in the spectral bandpass are essentially free of interfering absorption. H_2O absorbs near the adjacent lines.
HF	3 intervals, each passes a line between 4000 and 4080 cm^{-1}	3.0	PbS 0.0001		Intervals in spectral bandpass are essentially free of interfering absorption. H_2O interferes in P branch at lower wavenumbers.

it has been possible to reduce the interference by other gases to a negligible amount while measuring CO, SO_2, HCl, or HF.

It is anticipated that field tests performed with this versatile instrument will provide information that is helpful in the design of simpler gas cell instruments for a single gas species. Significant improvements in the performance of instruments designed for specific applications can undoubtedly be realized.

ACKNOWLEDGMENT

The research work leading to the instrument described in this paper was performed on an independent research and development program funded by Philco-Ford, and certain patentable features were reduced to practice on that program. These features are the subject of certain invention disclosures and patent applications currently in process by Philco-Ford. In addition, certain key aspects of the operation of the instrument are covered by Philco-Ford Patent No. 3,793,525. The instrument was designed and constructed in its present form for adaptation to across-the-stack measurements by Aeronutronic Division of Philco-Ford under Contract 68-02-0575 with the Environmental Protection Agency.

REFERENCES

1. Burch, D. E. and J. D. Pembrook. "Instrument to Monitor CH_4, CO, and CO_2 in Auto Exhaust," prepared by Philco-Ford Corp. for EPA under Contract No. 68-02-0587. EPA Report No. 650/2-73-030 (October, 1973).

2. Ludwig, C. B., R. Bartle, and M. Griggs. "Study of Air Pollutant Detection by Remote Sensors," prepared for National Aeronautics and Space Administration under Contract NAS12-630 by Convair Division of General Dynamics. Report No. GDC-DBE68-011 (December, 1968).

3. Bartle, E. R., S. Kaye, and E. A. Meckstrath. "An *In Situ* Monitor for HCl and HF," AIAA Paper No. 71-1049, presented at the Joint Conference on Sensing of Environmental Pollutants, Palo Alto, California (November, 1971).

4. Goody, R. "Cross-Correlating Spectrometer," *J. Opt. Soc. Am.*, *58*, 900 (1968).

5. Hanst, P. L. "Spectroscopic Methods for Air Pollution Measurements," in *Advances in Environmental Science and Technology*, J. N. Pitts and R. Metcalf, Eds. (New York: John Wiley and Sons, 1971).

6. Burch, D. E. "Adjustable Bandpass Filter Employing a Prism," *Appl. Opt.*, *8*, 649 (1969).

7. Decker, J. A., Jr. "Experimental Realization of the Multiplex Advantage with a Hadamard-Transform Spectrometer," *Appl. Opt.*, *10*, 510 (1971).

8. Abels, L. L. and J. H. Shaw. "Widths and Strengths of Vibration-Rotation Lines in the Fundamental Band of Nitric Oxide," *J. Mol. Spectrosc., 20,* 11 (1966).

9. Shaw, J. H. "Nitric Oxide Fundamental," *J. Chem. Phys. 24,* 399 (1956).

10. Burch, D. E., J. D. Pembrook, and D. A. Gryvnak. "Absorption and Emission by SO_2 between 1050 and 1400 cm^{-1} (9.5-7.1μm)," prepared by Philco-Ford Corp. for EPA under Contract No. 68-02-0013. Philco-Ford Report No. U-4947 (ASTIA PB 203523) (July, 1971).

11. Benedict, W. S., R. Herman, G. E. Moore, and S. Silverman. "The Strengths, Widths, and Shapes of Lines in the Vibration-Rotation Bands of CO," *Astrophys. J., 135,* 227 (1962).

12. Benedict, W. S., R. Herman, G. E. Moore, and S. Silverman. "The Strengths, Widths, and Shapes of Infrared Lines II. The HCl Fundamental," *Can. J. Phys., 34,* 850 (1956).

13. Meredith, R. E. and N. F. Kent. "Line Strength Calculations for the $0 \rightarrow 1$, $0 \rightarrow 2$, $0 \rightarrow 3$, and $1 \rightarrow 2$ Vibration-Rotation Bands of Hydrogen Fluoride," prepared for the Advanced Res. Proj. Agency under Contract No. SD-91 by The University of Michigan. Report No. 4613-125-T (April, 1966).

14. Herget, W. F., W. E. Deeds, N. M. Gailar, R. J. Lovell, and A. H. Nielsen. "Infrared Spectrum of Hydrogen Fluoride: Line Positions and Line Shapes. Part II. Treatment of Data and Results," *J. Opt. Soc. Am., 52,* 1113 (1962).

CHAPTER 11

ABSORPTION SPECTROSCOPY APPLIED TO STATIONARY
SOURCE EMISSIONS MONITORING

H. C. Lord

In order to meet the requirements of the process
industries for heavy-duty reliable instrumentation,
an analytical system based upon absorption spectroscopy
has been developed. By mounting the hardware directly
on the stack, duct, or process stream, and passing
the light beam through the gas stream, this system
provides both a continuous and instantaneous, as well
as integrated analysis of the gases desired. Coupling
this *in situ* mounting with a well-designed analytical
capability that provides a specific analysis, the
sample handling and conditioning system can be eliminated. Potentially this results in both a more
accurate and trouble-free operation.
     At the present time instrumentation is monitoring
NO, SO_2, CO, CO_2, hydrocarbons, and opacity in a
variety of locations including gas, oil, and coal-
fired boilers, steel mills, and refractories. Equipment will be going into both refineries and pulp and
paper mills in the near future. Data are presented
from some of these typical applications.

INTRODUCTION

     Light from a polychromatic source is collected
with a collimating mirror in the source box. This
collimated beam passes through the window interface
between the source box and gas stream, through the
gas stream, through the window interface between the

gas stream and the analyzer box, and is then collected and focused by the focusing mirror onto the analyzer module. This results in an integrated reading across the gas sample. This *in situ* monitoring eliminates the sampling system common to most other monitoring systems, thus minimizing the maintenance and down-time typically associated with these systems.

The analyzer module consists of a means to separate these wavelengths. Both a specific wavelength where the gas uniquely absorbs and an adjacent nonabsorbing wavelength are alternately measured with a single detector and then ratioed, resulting in a signal dependent only upon gas concentration. There is a separate detector for each gas, allowing simultaneous and continuous monitoring of each parameter.

Under normal operations, the gas analysis is in terms of volume concentrations, either per cent or parts per million. A measurement in a stack will be on a wet basis with some varying O_2 and CO_2, depending upon the process and also upon dilution, such as leakage. By measuring CO_2 in a boiler stack simultaneously with the desired NO or SO_2 emissions and comparing this with the predicted value for the fuel burned, the dilution at the monitored location is readily determined. This comparison can be done automatically, so that the emission data can be presented on a standardized basis such as 3% O_2 or 12% CO_2. Mass flow data are readily available, if desired. Since absorption is a direct function of the number of molecules present, the read-out can be calibrated either in concentration (ppm) or the instantaneous mass emission (lb).

INSTRUMENTATION

In situ monitoring in an industrial environment requires a rugged, reliable, simple, and yet specific system. Absorption spectroscopy gives this, plus versatility. Every gas of general interest, except O_2, can be measured directly in the near ultraviolet, visible, or near infrared. We use one or more sources of polychromatic radiation, one for each spectral region. The light is collimated, using two-inch square mirrors, sent through the window interfaces and through the sample, and then focused onto the entrance slit of the analyzer module (see Figure 11.1). The path length is defined by a slot

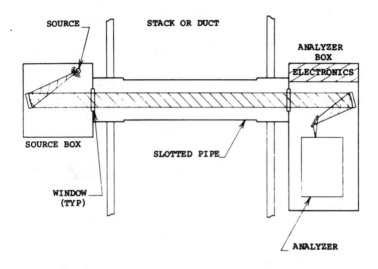

Figure 11.1. Optical layout of instrument.

in the pipe connecting source and analyzer boxes. A clean air purge is used to maintain the relative cleanliness of the window interface between the analytical system and the stack, and also to purge the pipe up to the slot. The shear with the lower pressure, higher velocity flue gas insures that the optical path length is known and constant. Details of the instrument-stack interface are shown in Figure 11.2.

A single beam from each source is used, and that beam is separated into its component wavelengths. Through prior laboratory work, we have selected a wavelength where the gas in question absorbs. This selection is based upon the concentration ranges required, the optical path length available, and freedom of absorption by potentially interfering species. The last parameter can usually be readily avoided by using moderately high resolution. Both this absorption wavelength and an adjacent non-absorbing wavelength are alternately measured by the detector (see Figure 11.2). A separate detector is used for each gas being measured, thus allowing simultaneous operation.

By measuring the adjacent nonabsorbing wavelength we have a continuous background reading. The ratio of I to I_0 is independent of any system or operational

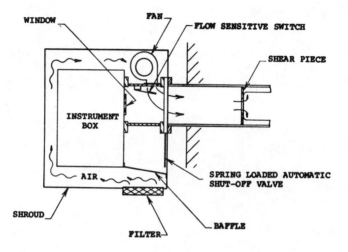

Figure 11.2. Details of stack-instrument interface.

changes, including source or detector decay, window
darkening, scattering in the gas stream due to particulates or water drops, or voltage change, and is
a direct measure of the concentration of gas being
measured, since from Beer's Law: $I = I_0 e^{-\alpha c \ell}$, where
α (absorption coefficient) is measured in the laboratory, and ℓ is fixed by the slot length.

A variety of analytical techniques are utilized,
including dispersive absorption spectroscopy,
utilizing a high resolution grating and narrow slits.
A variant of this would be correlation spectroscopy
wherein a series of slits positioned at wavelengths
where the sample absorbs (I) are ratioed with a
series for interweaving wavelengths where the sample
does not absorb (I_0). Gas cell correlation is also
employed, wherein a narrow band pass filter defines
the $\Delta\lambda$ being studied, and then a cell containing a
very high concentration of the gas is alternately
cycled in and out of the beam. When in the beam, all
the wavelengths where the gas absorbs are removed,
leaving only the nonabsorbing wavelengths, I_0. When
out, both I and I_0 are measured by the detector.
Finally, for certain applications, to increase signal-
to-noise, or to remove the effect of a rapidly
changing continuum intensity with wavelength,
derivative spectroscopy is used. By using a combination of these techniques, it has been possible

to measure each gas desired, in the presence of any
other gases, with no interferences.

Calibration of this equipment is performed in
two different ways. Since absorption of radiation
is directly proportional to the number of molecules
in the light beam as long as the total pressure and
temperature are the same, one can simulate 10 feet
of stack sampling path length containing 500 ppm of
NO by using 5% NO in 0.1 feet, backfilled with N_2
to 1 atmosphere pressure. Sealed cells of gas, containing the concentrations of gases desired and
prepared daily, are then placed in the collimated
light beam for the instrument calibration. The
analyzer module alone is checked by utilizing a
separate light source on a "zero-jig." The entire
system is then checked by using the instrument source
on the other side of the gas stream. If the plant
is operating, one cannot determine an absolute zero,
but rather reads some emission level. However, due
to the nonlinearity of the absorption curve, zero
and full scale can be readily determined by forcing
the addition of 100, 200, 300 ppm's of gas, etc.,
to add incrementally across the stack. In this way
a dynamic on-stream calibration is performed that
provides a check on the entire analytical system,
and is a calibration in the presence of the sample
and all of the other stack constituents.

APPLICATIONS

This equipment has been designed to measure up
to five parameters simultaneously with the same
installation. Gases being monitored presently include NO, SO_2, CO, CO_2, and hydrocarbons, as well
as opacity. Monitoring for NO_2, H_2S, SO_3, HCl, and
Cl_2 will be available in the near future. Potential
applications for this equipment include power plants,
steel mills, chemical plants, refineries, acid plants,
smelters, and pulp and paper mills. Installations
in stacks, ducts, or process streams are common.
Figures 11.3 and 11.4 show photographs of installations of this type in a power plant.

Table 11.1 shows the results of the comparison
of our integrated stack reading with an averaged
3-point traversal PDS test (Environmental Protection
Agency, Method 7, phenyldisulfonic acid procedure).
There is a very good correlation of the data.
However, Table 11.2 provides a closer look at

Figure 11.3.   Typical duct installation.

Figure 11.4.   Typical stack installation.

Table 11.1

Boiler Emissions, Oil Fuel

Load (Megawatts)	Boiler O_2 (%)	CO_2 (%)	O_2 (%)	H_2O (%)	Stack NO_x (ppm)	E.D.C. Stack NO_x (ppm)
196	3.8/3.9	10.7	6.7	10.4	201	200
183	4.0/4.0	10.3	7.2	8.3	171	175
215	4.3/1.2	10.7	6.5	8.5	203	200
210	4.5/2.5	11.0	6.4	9.5	219	210

Table 11.2

Boiler Flue Gas Point Analysis

		Dry NO_x (ppm)	O_2 (%)	NO_x (3% O_2, Dry)
Point 1	A	173	8.2	243
	B	179	7.8	244
	C	174	7.7	246
Point 2	A	221	6.8	280
	B	219	8.0	303
	C	216	7.2	282
Point 3	A	172	7.8	234
	B	183	7.9	251
	C	188	6.8	238
			Average	258 ppm

Environmental Data Corporation @ 3% O_2, dry, 5% NO_2				253 ppm

Load:	195 Megawatts	Fuel:	Oil
Boiler O_2:	East = 7.5%	West = 3.5%	

the individual PDS tests, and shows that considerable variation occurs in both space and time. In this case it was found that three points were sufficient to give good correlation to the integrated analysis.

DATA PRESENTATION

The most basic mode of presentation of the data is a continuous record on a strip-chart recorder located in the control room (see Figure 11.5). The data are presented in terms of volume concentrations, either per cent or parts per million. Time is moving right to left with a chart speed of two inches per hour. This shows a change in the ratio of low and high sulfur fuels, where both NO and SO_2 decreased There was a temporary upset in the hydrocarbon reading (usually ~ 0) at the beginning of the fuel change. Also note the large momentary increase in CO during the fuel change.

Most of the previous analytical methods are unable to measure gas concentrations under the existing conditions in the stack. In particular, since water may be an interference, many analytical techniques require removal of condensable water vapor prior to analysis. This analytical result (dry basis) will be higher than the actual emission conditions (wet basis, where water is a diluent).

Leakage in the system, such as at the air pre-heater and in the ducts following it, as well as variations in excess air, leads to a variable dilution of the sample. This dilution will affect equally all the pollutant emissions. The oxygen content would be expected to increase greatly, since the diluting gas (air) has a higher O_2 content than does the flue gas. This variable dilution in utility flue gas streams is most readily determined by measuring the carbon dioxide concentration in the stack gas and comparing it with the predicted value for the fuel being burned, based upon the H/C value. The water vapor content of the flue gas can also be calculated from a knowledge of the fuel (and its carbon-to-hydrogen ratio). The water vapor will also be diluted by air leakage, just as all the other gases are.

It may be important at some time to have the data available on a standardized dilution basis. Normally this is done by stating the emissions at either a 12% CO_2 level or a 3% O_2 level. The

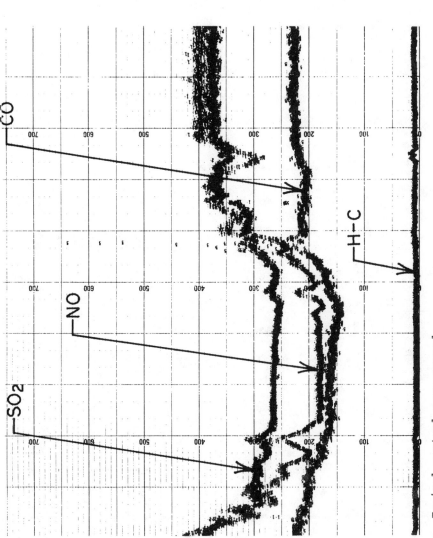

Figure 11.5. Typical control room record.

standardized emission values are determined in the
following way:

$$\text{Standardized Concentration of Sample} = \text{Measured Concentration of Sample} \frac{(12\% \; CO_2)}{(\text{actual} \; \% \; CO_2)}$$

OR

$$\text{Standardized Concentration of Sample} = \text{Measured Concentration of Sample} \frac{(21\% - 3\% \; O_2)}{(21\% - \text{actual} \; \% \; O_2)}$$

Either of these readings can now be corrected to a
dry basis as follows:

$$\text{Sample Conc. (Dry Basis)} = \text{Sample Conc. (Wet)} \left(\frac{100}{100 - \% \; \text{water vapor}} \right)$$

These manipulations of the emission data can be performed by the plant personnel, but if important on a
routine basis, they can be performed automatically
by simple circuitry.

Mass flow data are readily available, if desired.
Since this instrumentation measures absorption per
molecule, the read-out can be calibrated either in
concentration (ppm) or the instantaneous mass emission (lb). CO_2 analysis provides a simple means
of determining the required flow rate. The fuel
flow meter continuously provides a measurement of
the fuel usage (barrels/hour, . . .). From stoichio-
metric calculations and the carbon-hydrogen ratio of
the fuel, the CO_2 mass formation rate in the boiler
is determined.

The mass flow analysis is then derived by taking
the ratio of the calculated CO_2 emission in lb/hr,
from the fuel flow, to the measured instantaneous
CO_2 emission, and multiplying by the instantaneous
emission of the desired species, such as:

$$NO \; (\text{lb} \; /\text{hr}) = CO_2 \; (\text{lb/hr, fuel flow}) \frac{NO \; (\text{conc. or mass})}{CO_2 \; (\text{conc. or mass})}$$

Alternately, one can utilize the power generation
rate, in millions of Btu/hr, and provide the mass
emission rate as follows:

$$\text{lb. } NO/\text{million Btu} = \frac{\text{lb } NO/\text{hr}}{\text{millions Btu/hr}}$$

The simple circuitry that would automatically generate
the mass emission data is presently available.

SUMMARY

Absorption spectroscopy has been shown to be
capable of allowing continuous on-stream monitoring
of selected gases, providing an instantaneous response
as well as an integrated analysis of the gas stream.
By eliminating the sampling system, the hardware is
simplified, and maintenance is reduced to the analyzers
only. By measuring the absorption at a wavelength
where only the gas desired absorbs, and ratioing
this to a closely-spaced nonabsorbing wavelength,
using the same optical path and detector, the system
is continuously self-calibrated for light level
changes.
Absorption spectroscopy is a versatile analytical
technique, which can be implemented in several ways,
including dispersive, gas cell correlation, and
derivative spectroscopy. These allow the monitoring
of any gas that shows a unique absorption spectra,
and includes most of the common environmental pollutants as well as process control parameters.
Development work to increase the gas monitoring
capabilities is continuing, thus broadening the
utility of this technology by industry.

CHAPTER 12

REMOTE SENSING OF SO$_2$ IN POWER PLANT PLUMES
USING ULTRAVIOLET ABSORPTION AND INFRARED
EMISSION SPECTROSCOPY

H. M. Barnes, Jr., W. F. Herget and R. Rollins

INTRODUCTION

One of the important programs of the Chemistry
and Physics Laboratory is the investigation of remote
sensing instrumentation and methodology for stationary
source emissions. By the term "remote sensing" we
mean an electro-optical technique in which the analyzer
is physically removed (usually 100-1000 meters) from
the source being investigated.

This paper will concern itself with two such remote sensing instruments, one a commercially available
design based on dispersive correlation spectroscopy,
the other a research instrument utilizing conventional
dispersive grating techniques. Both of these instruments can be used in either the active (double-ended
with an active source) or passive (single-ended with
environmental background or stack gases serving as
the source of energy) mode. This paper will concern
their applicabilities to the passive mode only and
will deal with their uses on stationary sources and,
in particular, a coal-fired power plant.

245

CORRELATION SPECTROMETER SYSTEM

Description of the Instrument

The instrument shown in Figure 12.1 is the
Barringer COSPEC II, commercially available from
Environmental Measurements, Inc.* of San Francisco,
California. It consists of two telescopes to collect

Figure 12.1.   The Barringer COSPEC II at field site.

light from a distant source (usually solar radiation)
a two-grating monochromator for light dispersion, a
disc-shaped multislit mask, and an electronics system
for signal processing. The disc is composed of
circular slits photo-etched in aluminum on quartz,
and provides a high contrast reference spectrum for
matching against the incoming absorption spectra.
The arrangement is designed to correlate successively
in a positive and a negative sense with the SO_2
absorption bands via disc rotation in the exit plane
of the monochromator. The detector used is the XP
1118 photomultiplier tube. Signal processing results
in a voltage output proportional to the optical depth
in ppm-m of gas being observed.

*Mention of a company name or product is not intended to
constitute endorsement by EPA.

The dynamic range of the instrument is 100-10,000 ppm-m. Calibration is effected using four fused-silica cells containing known concentrations of SO_2 that are incorporated into the instrument. The instrument is 50 x 25 x 37 cm in size, weighs approximately 18 kg and can be operated by either a 12V DC battery or conventional 115V 60 Hz.

Description of the Field Testing Site

The data that will be discussed for the COSPEC II were obtained at the Duke Power Company River Bend Steam Station in Charlotte, North Carolina. The River Bend Station is a pulverized coal-fired power plant having eight stacks in a row approximately 6-8 meters apart. The stacks are 20 meters high, 2 meters in diameter, and exit through the roof of the seven-story main plant building. EPA personnel have instrumented stack #3 with SO_2 continuous monitors, a transmissometer, and a beta gauge for particulate mass monitoring. During the study the authors worked closely with other EPA personnel involved with the in-stack SO_2 measurements to compare their results with the remote measurements. The in-stack measurements referred to are described in detail in the paper by J. B. Homolya appearing elsewhere in this volume.

The proximate stack locations (Figure 12.2) presented serious problems in the remote measurements study with the COSPEC II. Separation of plume contributions from adjacent stacks was difficult under anything less than ideal conditions; the most important factor here seems to be a combination of stack gas exit velocity and wind speed. If the exit velocity is sufficiently high in relation to wind speed, then the plumes rise vertically for the necessary three or four stack diameters to isolate single plumes. This problem is one that all remote sensing techniques will have to contend with regardless of the measurement method.

Discussion of the Data

Figure 12.3 is a representative section of the strip chart data obtained with the COSPEC II. The bottom trace includes the calibration of the system. The upper trace is the AGC reading, a qualitative measure of the light input to the PM tubes. The

Figure 12.2.   Ground level view of River Bend steam station.

calibration cells are sequentially introduced to
give the signal level readings shown here; for this
work the dynamic range required only three calibration points for stack #3 at Charlotte. There are
eight combinations of cells possible to extend the
range to 10,000 ppm-m.
     The studies to be discussed here involved three
weeks of activity over a five-week period at the
plant working at three different ranges from 20
meters to 300 meters slant range.
     Table 12.1 shows the data taken at the three
ranges. From the table one sees that agreement with
the in-stack readings at the closest range is quite
good. However, at greater distances correlation
becomes quite poor. The value at 200 m is suspect
since it represents only one hour of data taken on
one day. On that day the plant experienced an out-
age terminating further work during our study. At
300 m the results are low but reasonably consistent
(52 ppm *vs*. 60 ppm). This consistency may be
applicable in using the instrument at extended ranges
if the range correction factor is constant under
different meteorological conditions, such as rain,
haze, and smog. Much more work would be needed before
a firm conclusion could be reached on this matter.

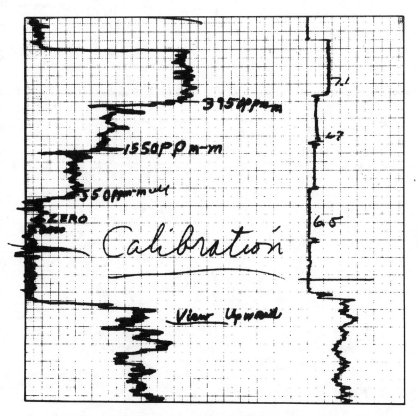

Figure 12.3.   Stripchart record of COSPEC II calibration.

Table 12.1

Comparison of In-Stack and Remote
SO_2 Concentration Measurements

Range (meters)	In-Stack (ppm)	Remote (ppm)	$\dfrac{(IS) - (REM)}{(IS)} \times 100$
20	500	465	7
20	585	520	10
200	700	240	65
300	460	224	52
300	450	205	54
300	700	275	60

However, it should be noted that signal loss results
in a sensitivity degradation ultimately affecting
the useful slant range and dynamic limits of
concentration.

A more serious problem is represented in Figure
12.4. With the cooperation of the power plant personnel, the banks of electrostatic precipitators on
the stack were sequentially inactivated, thus varying

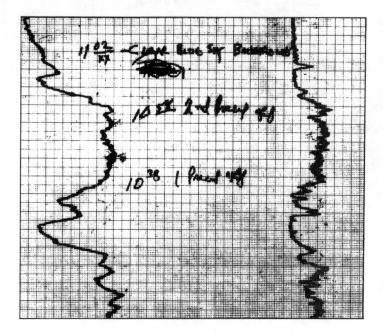

Figure 12.4. Effect of particulate loading on COSPEC II signal.

the plume opacity. When the second bank of precipitators was turned off, resulting in a 20% opacity
reading (by in-stack transmissometer), the COSPEC II
signal level dropped to an unmeasurable level although
the in-stack SO_2 monitors showed no change in SO_2
readings. Since this level (20% opacity) is within
federal regulations, the problem limits the usefulness
of the instrument severely.

The particulates problem can be represented in
Figure 12.5. Light reaching the sensor includes
scattered light in and out of the plume; it also
includes light in and out of the field of view. In
Figure 12.5 rays a, b, c represent radiance from the

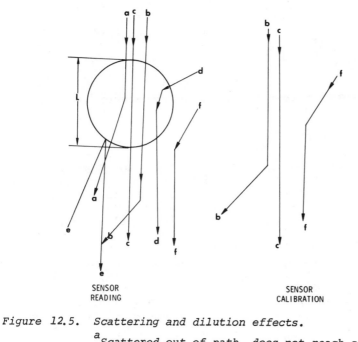

Figure 12.5. Scattering and dilution effects.

$_a$Scattered out of path, does not reach sensor
$_b$Scattered by intervening atmosphere
$_c$Desired light ray
$_d$Incomplete signature of plume absorption
$_e$No plume signal, scattered off front face of plume
$_f$Pure dilution light.

background that is incident on the plume backside
and that is in the optical path of the sensor. Ray
a is scattered out of the optical path by plume
particulates; ray b is also scattered out of the
optical path, but by the intervening atmosphere;
ray c is the desired radiation that traverses the
entire plume and reaches the sensor. Rays d, e, and
f are scattered into the field of view, but do not
contain the total signature from the plume. In
particular, ray f could be called "pure dilution
light" since it contains no part of the plume
signature at all.

It appears that the AGC compensation is inadequate
to offset scattering both by the intervening atmosphere and plume particulates.

One other problem encountered during the field
tests was that of the plume not remaining in the
field of view of the COSPEC II. This occurred even
on moderately windy days. The COSPEC II FOV is 3 mr
x 10 mr. Reducing that to 3 mr x 3 mr or perhaps
3 mn x 5 mr should allow operation during moderately
windy days when the plume is not reasonably vertical
for more than one or two stack diameters.

Summary and Conclusions

In conclusion, the COSPEC II in its present
design is inadequate for EPA use for routine monitori
purposes. The main limitation, as has been noted,
is inadequate AGC compensation for scattering effects
by particulates both in the plume and in the intervening atmosphere. The less serious problem of field
of view is perhaps a more readily rectified one.
It should be noted here that EMI, Inc. recommends
usage of the COSPEC in two other modes, namely as a
mobile perimeter monitor (looking vertically into the
sky) and from helicopters (sighting horizontally
through plumes). The authors are planning a field
test program of the mobile monitoring mode and will
report on this in the future.

DISPERSIVE SPECTROMETER SYSTEM

Description of the Instrument

The dispersive spectrometer system, which bears
the acronym ROSE (Remote Optical Sensing of Emissions
was designed and fabricated for EPA by General
Dynamics/Convair, San Diego, California. The contractor's final report[1] describes the ROSE system
in complete detail; the major system components will
be reviewed here. Figure 12.6 shows the principal
parts of the receiver section: Dall-Kirkham telescopic optics with a 60-cm diameter primary mirror,
Perkin-Elmer Model 210 linear wave number drive
monochromator, and detector housing with a portion
of a Cryogenic Technology closed-cycle helium
refrigeration system. The monochromator contains
two gratings that are each used in first order with
long-wave pass filters to cover the spectral ranges
2.8-6.5 and 6.5-14 microns. The Ge:Hg detector is

Figure 12.6. ROSE system receiver section.

operated at 28°K. The source unit (not shown) contains identical telescopic optics, and a blackbody
radiation source is located at the telescope focus.
The blackbody is operable over the range 1100–1800°K
and is optically chopped for long-path (≤ 2 km)
absorption measurements. In the remote emission
measurements described herein, optical chopping
occurs just in front of the monochromator entrance
slit. For intensity calibration purposes the source
unit is placed just in front of the receiver unit
so that the blackbody radiation (unchopped) fills
the field of view of the receiver unit. Both source
and receiver units contain bore-sighting attachments
to allow visual alignment of the telescopes.
    Figure 12.7 shows the electronic consoles for
the system. The left console contains phase-sensitive
detection and amplification circuits, wave number
drive controls, and a strip chart recorder. The
other console contains analog-to-digital conversion
electronics, digital voltmeters for visual displays,
and components for printed paper read-out and magnetic tape recording. In all forms of data display,
the principal information is signal intensity as a
function of wave number.

Figure 12.7.   ROSE system electronic consoles.

Description of Field Test Sites

Spectral emission data were obtained at two sites with the ROSE system. The Duke Power Company site in Charlotte, North Carolina, was described previously. The second site was a pulverized coal-fired power plant in Asheville, North Carolina, operated by the Carolina Power and Light Company. The advantage of the Asheville site was that during most of the measurements only one of their two stacks was in use and it was therefore possible to obtain single plume spectra as desired. The disadvantages of the Asheville site were that the stack wall is essentially at ambient temperature (making it difficult to measure atmospheric attenuation between the plume and the receiver) and the in-stack measurements were not as extensive as those at Charlotte.

In addition to the extensive in-stack measurements at the Charlotte site, the multiplicity of stacks allow, with selection of the proper wind conditions, collection of data with a variety of plume-background combinations. The distances from the receiver unit to the stack exits were 650 meters at an elevation

angle of 25° at Asheville and 400 meters at an
elevation angle of 10° at Charlotte. The Charlotte
site contains four oil-fired gas turbine units used
during peak load periods. A few spectra of the
effluent from these units were obtained along a
horizontal line-of-sight. Although the ROSE system
can operate off generator power, shoreline power
was used at each site.

Data Calibration, Collection,
and Reduction Techniques

The emission spectra are calibrated in terms of
wave number and intensity. Wave number calibration
is obtained by analysis of absorption spectra of
common gases. Occasional checks of the original
calibration curves furnished by the contractor have
shown them to have remained essentially unchanged.
Intensity calibration is made with the source blackbody unit in the laboratory, as described above,
before and after a field measurements trip. The
intensity calibration is made in spectral radiance.
Since signal voltage is proportional to spectral
radiance, a comparison of the signal produced by the
stack plume to that produced by the blackbody (at a
known temperature) gives plume spectral radiance for
any desired wavelength. Since the lowest temperature
of calibrated operation of the source blackbody is
$\sim$1100°K and the plume temperatures were $\sim$400°K, it
was necessary to record the blackbody spectrum with
narrower slits and less amplification than used for
the plume. Overall system linearity and lack of
dependence on spectral resolution for the blackbody
spectrum allowed these factors to be taken into
account. A low temperature blackbody has been
obtained to simplify the intensity calibration procedure in the future. A reference light source
internal to the receiver unit is used to verify
that no degradation of performance of the electronic
and optical components occurs during transportation.
The data collection procedure itself is relatively simple. The monochromator slit width, the
various signal processing controls, and the spectral
scan speed are determined so as to give maximum
spectral resolution consistent with reasonable
signal-to-noise ratio. Generally, four types of
data are obtained for each spectral region:

1. The receiver is sighted on the plume (the resulting signal will include the effects of
   atmospheric emission on the far side of the
   plume, the emission of the plume itself, and
   attenuation of signal between the plume and
   the receiver).
2. The receiver is sighted off to the side of the
   plume (the resulting signal will be due to the
   sky background).
3. The receiver is sighted on the stack wall just
   below the stack exit. (If the stack wall is
   significantly warmer than ambient, the spectrum
   will allow an estimate of the atmospheric
   attenuation between the plume and the receiver.)
4. The receiver is sighted on the plume and observations are made at selected constant wavelengths for several minutes. (The signal gives
   a measure of the temporal fluctuations in the
   plume.)

The spectral scans described in (1), (2), and (3)
are generally repeated several times to verify which
fluctuations in the signal are spectral rather than
temporal. For all spectra shown below, the resolution was 10 cm^{-1} in the 4-6μ range and 4 cm^{-1} in the
7-13μ range; the scan time was about 3 minutes for
each spectral range.

Data were taken at the Asheville site during day
and night conditions and when the plant was known to
be under different operating conditions. Data at
the Charlotte site were taken under varying wind
conditions, which produced different background conditions in addition to the usual sky background. It
was possible to obtain data on a particular plume
with dispersed dirty plumes as part of the background
and also with dispersed clean plumes as part of the
background.

Reduction of passive infrared emission spectral
data to obtain species concentrations can be about
as complicated as one desires to make it. A complete
computerized data reduction program would require
the capability of analyzing the plume emission
spectra and the background sky spectra in order to
obtain the plume temperature and then calculate plume
species concentrations. This type of program would
generally require reiterative calculations using the
basic molecular parameters of all the species involved
to calculate spectra which, when the correct temperatures and concentrations are finally determined, would
match the measured spectra. This type of data reduction procedure may eventually become commonplace, but

it is certainly not considered practical at this
time. (This type of procedure is practical in a
double-ended absorption measurement where temperature
can be determined in a simple manner.)

The basic purpose of the ROSE system is to study
the spectra of various stationary source emissions
so as to determine which spectral regions are most
suitable for measurement of particular species and
to verify which species are actually present in the
effluent. This information can then be used to assist
in the design of simple instruments for specific
pollutants. Thus, the type of data reduction
described above was not attempted. However, since
the plume temperature was measured during the course
of the in-stack measurements, it was possible to
calculate SO_2 concentrations from the observed spectra
with relative simplicity, and to allow an evaluation
to be made of the remote sensing technique under
ideal conditions.

The equations describing the thermal radiation
produced by a gaseous species in the plume are as
follows:

$$\varepsilon_p(\lambda) = \frac{N_p(\lambda)}{N_{BB}(T_p, \lambda)}$$

and

$$1 - \varepsilon_p(\lambda) = -K(\lambda, T_p)\, CL$$

where

$\varepsilon_p(\lambda)$ = the plume emissivity at wavelength λ

$N_p(\lambda)$ = the plume spectral radiance at λ (watts cm^{-2}
steradian^{-1} micron^{-1})

$N_{BB}(T_p, \lambda)$ = the spectral radiance of a blackbody at the
plume temperature, T_p, at λ

$K(\lambda, T_p)$ = the spectral absorption coefficient of the
species in question at T_p and λ

C = species concentration (molecules/cm^3)

L = optical pathlength through the plume

The quantity $N_p(\lambda)$ is determined from the spectra as
discussed under calibration procedures. Since the
plume temperature was measured directly, $N_{BB}(T_p, \lambda)$
could be calculated, and thus, the plume emissivity
was determined. The measurement wavelength λ was

selected so as to minimize the effects of atmospheric
absorption.   The plume optical path was taken as the
stack exit diameter.   The plume completely filled
the field of view of the instrument (approximately
0.1 x 1.0 meters with long dimension vertical).
Desired values of $K(\lambda, T_p)$ were taken from the work
of Burch.[2]   The above data then allowed calculation
of the SO_2 concentration.

Results

     Figures 12.8 and 12.9 show typical spectra ob-
tained from a coal-burning power plant plume.   This
"clean" plume is low in particulate content (opacity
less than 3%).   Figure 12.8 covers the spectral range
from approximately 7-13.5 microns.   The lower trace
shows the sky background spectrum, which did not vary
from day to night, and which consists primarily of
O_3 and H_2O at atmospheric temperatures appearing as
emission lines over the cold sky continuum.   When
not looking at the sky (or plume), the detector sees
an ambient temperature blackbody, whose signal level
(zero reference) is indicated at the left and right
extremes of the spectrum.   The upper trace shows
plume emission and sky background together.   Emission
due to SO_2 and CO_2 is clearly evident.   The 8.7μ SO_2
band is well defined, but suffers from moderate
atmospheric interference.   Atmospheric H_2O interferes
strongly with the 7.35μ SO_2 band and would be useless
for a remote measurement.   The reason that two spikes
appear will be apparent in later figures.   Figure
12.9 shows similar spectra for the 4-6μ spectral
range.   Below 3.9 and beyond 5.5μ sky and plume-sky
spectra are identical.   The weak CO_2 bands on either
side of 5.0μ are partially attenuated by H_2O absorption between the plume and the receiver.   Atmospheric
CO_2 absorbs out the center of the warm CO_2 emission
from the plume at the 4.3μ band.   Of particular
interest is that the SO_2 band at 4μ, though weak
compared to the 7.35 and 8.7μ bands, suffers from
no molecular atmospheric interference.   The structure
observed in these spectra were all reproducible and
were not due to temporal variations.   Except during
high wind conditions, the temporal fluctuations were
only a few percentages of the signal level.
     Figure 12.10 shows long wavelength spectra ob-
tained by looking at the warm stack wall (lower curve)
and a "dirty" plume (upper curve, $\sim$100% opacity).
The stack wall spectrum shows H_2O attenuation between

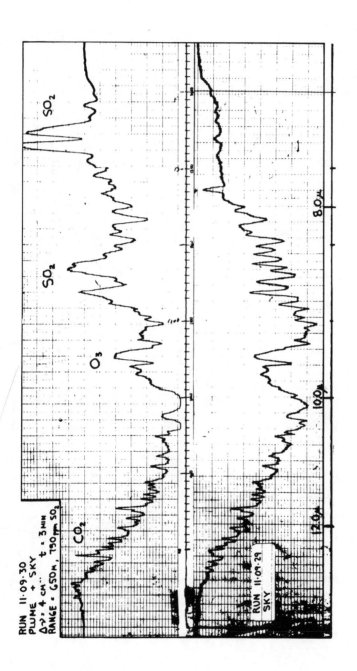

Figure 12.8.   Plume and sky spectra from 7-13 microns.

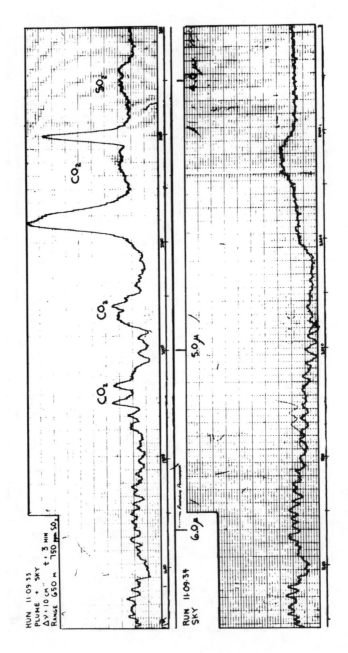

Figure 12.9.   Plume and sky spectra from 4-6 microns.

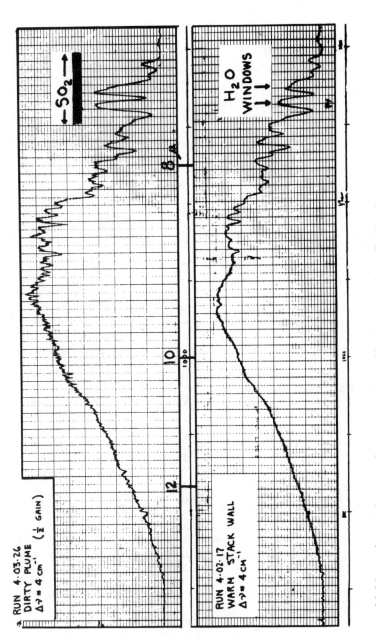

Figure 12.10. Dirty plume and warm stack wall spectra from 7-13 microns.

the stack and the receiver. The spectrum of the
dirty plume, which was recorded with half the amplification of the stack wall, shows continuum emission
due to fly ash; the spectral detail of the 8.7μ SO_2
band is washed out, but in the case of the 7.35μ
band, the two spikes remain. This indicates that
although the dirty plume has an emissivity of one
in the visible spectral range, it is partially
transparent in the infrared. In fact, using the
measured in-stack plume temperature, the dirty plume
emissivity was calculated to be 0.6 at 8.6μ. This
drop in emissivity for wavelengths larger than the
mean particle size is expected.[3] This fact could be
of particular importance in all types of optical
plume measurements, since the infrared could be used
to measure gas (SO_2 and NO, for example) concentrations in a stack whose particulate concentration was
too high to allow ultraviolet measurements. A high
frequency oscillation can be observed on the dirty
plume spectra. Constant wavelength scans show these
temporal fluctuations to be on the order of 5-10 Hz;
they are due presumably to density fluctuations.

The middle curve of Figure 12.11 shows the
spectrum of a clean plume with two dispersed dirty
plumes forming the background; the spectra of the
dispersed dirty plumes are shown in the lower curve.
It is seen that the dispersed dirty plumes form a
nearly opaque, ambient temperature screen behind the
clean plume, and that the plume spectrum is therefore
much less complicated. The upper spectrum was obtained from the turbine exhaust effluent produced
by combustion of air and kerosene. This spectrum
differs from the plume spectrum in that no SO_2 and
considerable H_2O was observed. The background
spectrum for the turbine exhaust was a continuum
produced by light ground fog, so that all details
in the spectrum were due to the exhaust itself.
All structure observed that is not due to CO_2 (as
indicated on the middle curve) is due to H_2O. A
comparison of the middle and upper curves in Figure
12.11 shows why the 7.35 micron SO_2 band appears as
two sharp spikes. There are two "windows" in the
atmospheric H_2O absorption spectrum at approximately
1328 and 1347 cm^{-1}; the center of this SO_2 band is
actually at 1360 cm^{-1}. The dispersed clean plume
had no measurable effect on the sky background.

Because of the freedom from atmospheric interference, the data shown in spectral runs 4-03-20 and
4-03-21 (Figure 12.11) obtained at the Charlotte
field site were reduced to obtain plume SO_2

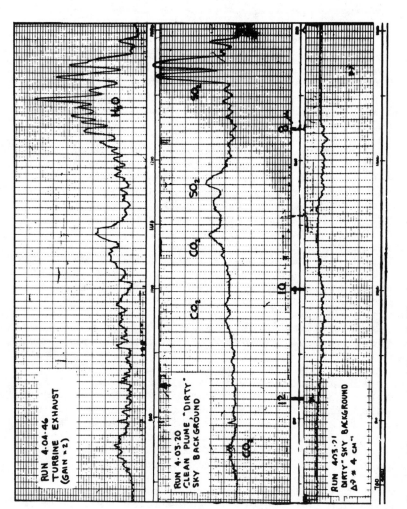
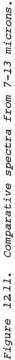

Figure 12.11. Comparative spectra from 7-13 microns.

concentrations. Since any practical remote sensing
device could automatically take into account atmospheric interferences, it was felt that these data
would give a good indication of the feasibility of
and problems involved in a remote sensing measurement, without the complications introduced by the
atmosphere. The wavelength selected for the concentration measurement was at the point of maximum
intensity in the R branch of the 8.7μ SO_2 band,
which is at 8.59μ (1165 cm^{-1}). The difference in
signal level between the above two runs (plume and
background) at this wavelength gave the signal level
due to SO_2. The calibration procedures described
earlier yielded a plume spectral radiance $N_p(8.59μ)$ =
4.83 x 10^{-4}. The measured in-stack plume temperature
was 414°K at the time of the spectral measurement.
The plume temperature at the line of sight of the
optical measurement (approximately 20 meters above
the point of the in-stack measurement) was taken to
be 400°K, which yielded a plume emissivity of 0.096.
Burch[2] has measured values of emissivity as a function of absorber thickness U (molecules cm^{-2}). (The
quantity U is the reciprocal of the spectral absorption coefficient $K(λ,T_p)$, and C = U/L.) From Burch's
curves a value of U = 0.25 x 10^{19} molecules cm^{-2} was
obtained. From these data a plume SO_2 concentration
of 650 ppm was calculated. At the time of the remote
measurement, an extractive monitor was reading 710
ppm and an in-stack optical instrument was reading
680 ppm.
   It is of interest to examine the sources of error
in the remote IR measurement. The blackbody source
used for intensity calibration is accurate to better
than 1% and any resulting error is less than 1%.
Temporal fluctuations in the plume signal can be
measured and averaged if necessary. The measurements
of Burch allow the quantity U to be determined to
within a few percentages, and for plume temperatures
on the order of 400°K and the wavelength selected,
U is relatively insensitive to temperature over a
range of ±25°K, so that temperature errors do not
strongly affect U. However, the quantity $N_{BB}(λ,T_p)$
is strongly dependent on temperature for the
temperature range under consideration, and this is
the most serious source of error in the measurement
technique. At T_p = 400°K, an error of ±25°K (±6%)
(∿±50°F) produces an error of ∿±20% in $N_{BB}(λ,T_p)$,
and the resulting error in C is ∿∓12% (*i.e.*, use of
a lower than actual temperature results in a higher
than actual concentration).

In hopes of finding a simple way of determining the plume temperature remotely, one additional calculation was made. Because of its optical depth in the plume, CO_2 will have an emissivity of 1 over the central portion of the strong 4.3μ band. Since the warm CO_2 in the plume emits over a wider spectral range than that over which the ambient CO_2 absorbs between the plume and the receiver (Figure 12.9), the possibility exists that a CO_2 emissivity of one (unattenuated by the atmosphere) could be observed in the region of 2250 cm^{-1}. If this were the case, the temperature could be determined directly, since the instrument would be observing blackbody radiation at that wavelength. Unfortunately, at ranges considered practical for routine remote monitoring measurements, *i.e.*, off the boundary of the source under observation, calculations showed that the region where ε_{CO_2} = 1 were obscured by the intervening atmospheric CO_2. It would generally not be practical to make an accurate range measurement to correct for the CO_2 absorption.

Summary and Conclusions

These above results may be summarized as follows. Under the idealized conditions in which atmospheric interference has been minimized and the plume temperature is known, a remote concentration measurement can be made with reasonable accuracy using dispersive spectroscopic techniques. For a practical measurement system the atmospheric interference problem can be handled reasonably well, but the temperature determination remains a problem. The most accurate measurement of T_p would require high spectral resolution (< 0.5 cm^{-1}) and a computerized data reduction program. It may be possible to use nondispersive techniques, such as interference filters, to isolate certain spectral regions, but the temperature calculation still requires a computer. What is needed, obviously, is a measurement technique that is relatively insensitive to temperature. Laser scattering techniques are not particularly temperature sensitive, but the cost of laser systems appears prohibitive for routine monitoring at the present time. One form of nondispersive infrared, called gas-cell (or gas-filter) correlation spectroscopy, appears to have the capability of making a single-ended remote infrared measurement which is relatively insensitive

to temperature. This technique is being studied by
JRB Associates and Philco-Ford Corporation under
contracts with EPA.

ACKNOWLEDGMENT

The authors wish to acknowledge the excellent cooperation
of Duke Power Company and the Carolina Power and Light Company
for allowing these measurements to be made at their facilities
and for their assistance provided during the measurements.

REFERENCES

1. Streiff, M. L. and C. L. Claysmith. *Design and Construc-
    tion of a System for Remote Optical Sensing of Emissions*,
    EPA-R2-72-052   (San Diego, Cal.: General Dynamics Corp.,
    1972).
2. Burch, D. E., J. D. Pembrook, and D. A. Gryvnak. *Absorp-
    tion and Emission by SO_2 between 1050 and 1400 cm^{-1}*
    U-4947 (Newport Beach, Cal.: Philco-Ford Corp., 1971).
3. Stull, V. R. and G. N. Plass. "Emissivity of Dispersed
    Carbon Particles," *J. Opt. Soc. Am., 50,* 121 (1960).

CHAPTER 13

A REVIEW OF AVAILABLE TECHNIQUES FOR COUPLING
CONTINUOUS GASEOUS POLLUTANT MONITORS
TO EMISSION SOURCES

James B. Homolya

INTRODUCTION

The state-of-the-art in the development of source
level pollution monitors has reached the stage at
which several viable detection methods are available.
Instrument systems have been designed that are based
on either process analyzer concepts or modification
of ambient air sensors. Measurement techniques such
as nondispersive infrared or ultraviolet absorption
have long been incorporated in the analysis of
process streams, whereas flame photometry, chemiluminescence, and electrochemical transducers have
found initial applications in ambient air monitoring
instrumentation where high sensitivity is required.
Nearly all of these sensors have the required
sensitivity, freedom from interferences, and adequate
response time where application for source level
pollution monitoring presents no problem on the
detectors themselves. But we find that for monitoring
stationary sources of gaseous emissions, consideration
must be given to the complete monitoring system.
Stack emissions usually contain corrosive gases at
elevated temperatures. Such streams may have a high
dew point temperature or include particulate matter
of varying composition and size. A monitoring system
must be capable of continuously extracting a sample
from these types of sources, transporting it to the
detector, and conditioning it, if necessary, for an
accurate analysis.

It is important that the extraction, transport,
and conditioning of the sample be consistent with
the analytical method involved. At present, there
are three sampling conditioning techniques available:
(a) a "brute-force" approach, (b) dilution techniques,
or (c) *in situ* measurement.

SOURCE LEVEL SAMPLE CONDITIONING

Figure 13.1 illustrates a typical measurement
system for extractive monitors having application
for the analysis of SO_2 or oxides of nitrogen from
combustion sources. The gas sample is withdrawn

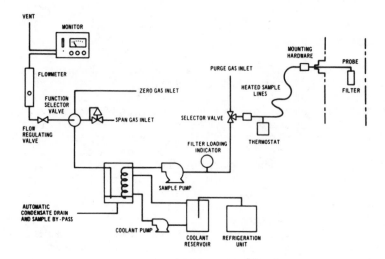

Figure 13.1.   Typical measurement system for combustion sources.

from the stack via a filtered-probe and passed through
a water removal system (usually a refrigerated dryer)
before entering the analyzer itself. For long term
operation, the water condensate is removed continually
and the probe filter is periodically back-flushed
with compressed air to remove entrained particulate
matter. The system also contains some provision
for the introduction of zero and calibration gases.
Several potential sources of error can exist in such
a sampling system prior to the instrument detector.
Sample integrity can be destroyed by: (1) chemical

reaction with surface materials, (2) chemisorption on particulate matter, (3) solution in the water condensate, and (4) leakage of sample lines. The Environmental Protection Agency has contracted an investigation into these problem areas.[1] Hopefully, design criteria can be established for analyzer-interface combinations applied to general stationary source categories.

An extractive monitoring system for combustion sources requiring minimal sample conditioning is illustrated in Figure 13.2. An air aspirator is

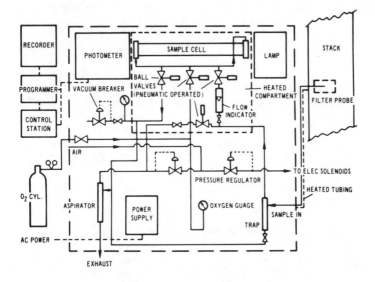

Figure 13.2. "Pumpless" monitoring system.

utilized to extract a sample from a source on a con-tinuous basis. Because our experience has demonstrated that sample pump failure has been a major problem in continuous source monitoring, the use of aspirators could be advantageous if a source of plant or instru-ment air is available at the system installation site. This particular system, incorporated as part of a research study discussed later in this paper, is the DuPont 460/1* SO_2 and NO_x analyzer. Stack gas is passed through a heated sample cell positioned between

---

*Mention of company name or product is not intended to constitute endorsement by EPA.

an ultraviolet energy source and a phototube detector.
Sulfur dioxide and the nitrogen oxides in the sample
gas are analyzed in sequence. A split-beam photometer
is utilized by measuring the difference in energy
absorption at 280-nm and 436-nm measuring wavelengths
and at a 570-nm reference wavelength. The 280-nm
wavelength is chosen for SO_2 measurement, with the
436-nm wavelength selected for NO_2 measurement.
Since nitric oxide has little absorbance in the
visible and ultraviolet, conversion to NO_2 is required
for its measurement. The system achieves this conversion by reacting NO in the sample with oxygen at
high pressure. By measuring the reaction product at
roughly 90% completion, a sequential analysis of
SO_2, NO_2, and NO can be accomplished every 15 minutes.
Between each analysis sequence the gas cell, heated
sample line, and probe filter are backflushed automatically with air to remove particulate matter from
the system and obtain a "zero" for the photometer.
    The particular design of an extractive monitoring
system depends largely on the characteristics of the
emission source. For example, Figure 13.3 illustrates
a typical monitoring system configuration for analyzing
the atmospheric emissions from sulfuric acid plants.

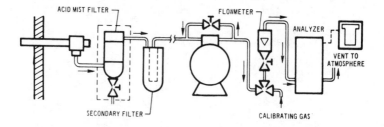

Figure 13.3. Sulfuric acid plant monitoring system.

In this arrangement, the probe filter has been
eliminated because of the absence of particulate
matter in the source emission stream. The filter
has been substituted with a coalescing device to
collect sulfuric acid mist before contaminating
the analyzer. This type of system has been widely
used in conjunction with many of the nondispersive
infrared analyzers.

SAMPLE DILUTION TECHNIQUES

Dilution techniques can offer an advantage in
sample conditioning by eliminating heated sample
lines and water vapor removal systems, if the stack
gas sample can be quantitatively diluted as close to
the source as possible. Such devices are based on
controlled flow, permeation sampling, or mechanical
means.

Figure 13.4 illustrates a controlled-flow dilution
system for monitoring stationary sources. The stack
gas sample is extracted via a filtered probe and
transported to the dilution network and analyzer by
heat-traced sample line. At this point, the source
sample is quantitatively diluted with air by a
controlled-flow/orifice combination. In this particular system, sulfur dioxide is measured by a
flame photometric detector. Principally, the dilution
concept is incorporated in this device to dilute
the SO_2 level into a range of linear detector
response. The dilution ratio is in the range of
1000:1. The sample stream must be filtered from
particulates to avoid altering the orifice dimensions
that would change the dilution ratio. In addition,
a constant temperature must be maintained at the
dilution network.

More recently, the controlled-flow approach has
been applied to an *in situ* dilution system.[2] In
this application, a specially-constructed sampling
probe acts as the orifice but with attainable dilution
ratios ranging from 2:1 to 20:1.

Diffusion or permeation sampling devices have
been widely reported in the literature.[3,4] A general
configuration for such systems is illustrated in
Figure 13.5. A source sample stream is introduced
into a chamber divided by a membrane permeable to
the gaseous component of interest. Gases permeating
the membrane are swept from the chamber by a carrier
stream and delivered to the analyzer. Both FEP
Teflon and silicone polymers have been used as membrane material for SO_2 and NO dilution. In addition,
polymer tubes have been substituted for the membrane.
In this manner, the tube is enclosed in a temperature-
controlled chamber and the sample stream passes over
the outer surfaces of the tube with carrier gas flowing
through the tube. The desired dilution ratio is
dependent upon: (1) the permeability of the membrane
or tube to the component of interest, (2) the surface
area of the membrane or tube and its temperature,
and (3) the volume flow of the carrier gas stream.

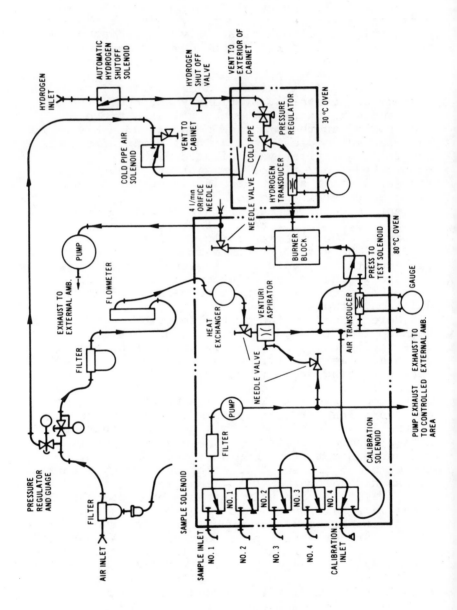

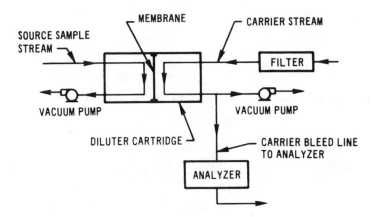

Figure 13.5. Diffusion diluter/analyzer network.

Systems utilizing this technique are commercially available from several manufacturers. However, the permeation samplers still require extraction of a source sample that must be filtered and held at an elevated temperature prior to entering the diffusion chamber.

Recently, a mechanical device has been developed in our laboratory to quantitatively dilute a source sample *in situ*, eliminating the need for heated sample lines and probe filters.[5] The system, illustrated in Figure 13.6, utilizes a rotating disc containing sample chambers of known volume. The disc is sandwiched between two stationary discs having sample inlet ports to allow gas exchange between the sample chambers and the stack gas environment. In practice, the dilution head is inserted into the emission stream. Rotation of the sample disc effects gas exchange at the sample inlet ports and at a mixing chamber into which a diluent gas is introduced. The diluted sample stream is then analyzed by an ambient air analyzer. The dilution ratio depends upon: (1) the number and volume of sample chambers, (2) the rotational speed of the sample disc, and (3) the volumetric flow of the diluent gas. Operation of the sampler has been demonstrated in the field by the continuous analysis of the SO_2 emissions from a 190-megawatt pulverized coal boiler.

Figure 13.7 represents a typical 24-hour segment from a week's continuous operation during which the "disc diluter" was coupled to a conductometric

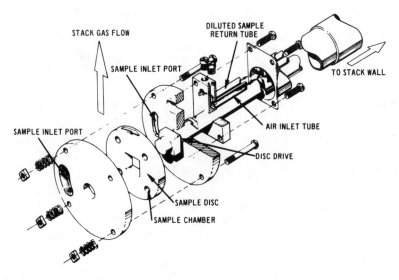

STACK GAS FLOW

DILUTED SAMPLE RETURN TUBE

SAMPLE INLET PORT

TO STACK WALL

SAMPLE INLET PORT

AIR INLET TUBE

DISC DRIVE

SAMPLE DISC

SAMPLE CHAMBER

Figure 13.6. Disc diluter.

ambient SO_2 monitor. The resultant SO_2 emissions, based on a dilution ratio of 1600:1, are plotted against the net load, in megawatts, from the boiler turbine generator. The diluter/analyzer combination appears to follow the trends in power output quite consistently. Further development of the dilution system is being carried out under contract to determine its range of applicability.[6] Also, design modifications have been incorporated to allow *in situ* calibration of the dilution head.

IN SITU MONITORING

In-stack measurement avoids any extraction of sample by utilizing the sample stream itself as an analysis chamber. These instrument systems employ electro-optical detection that can be arranged in three differing configurations.

A folded-path design places the energy source and receiver at the same location. In this manner the energy beam enters the emission stream through a slotted probe and is reflected back into the instrument. For large stack or duct diameters, the pathlength of measurements might be representative of a relatively small portion of the stack diameter.

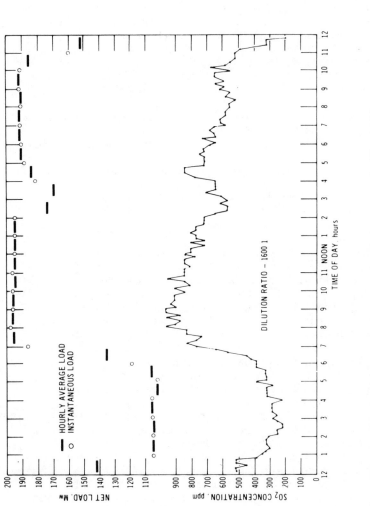

Figure 13.7. SO₂ concentration, ppm vs. net load, Mw, March 17, 1972.

A double-ended system is one in which the source and receiver are located at opposite ends of the stack diameter. However, some instruments still might require the use of a slotted pipe extending across the stack to either prevent misalignment of the optical beam or restrict the absorption path-length to maintain a linear detector response.

Recently, an investigation has been completed that compared both extractive and *in situ* electro-optical instrumentation for the measurement of SO_2 emissions from a pulverized-coal power generating boiler. An assessment was made of individual system performance under field conditions. To accomplish this, particular areas of interest in this study included:

1. an investigation of the effects of variations in fuel composition, boiler operating conditions, and particulate matter on the various measurement systems
2. a correlation of instrument response with standard EPA compliance test methods[7]
3. a determination of several instrument operating criteria such as zero drift, span drift, and maintenance.

This study was carried out at the Duke Power Company's River Bend Steam Station in Charlotte, North Carolina, from January through March, 1973. Table 13.1 outlines the instrumentation utilized during the study. Sulfur dioxide levels were simultaneously monitored by three discrete systems. They consisted of the DuPont 460/1 source monitoring system for SO_2, NO, and NO_x, the CEA Mart IV *in situ* SO_2 system, and the Bailey Meter Company SO_2 source analyzer. An in-stack transmissometer measuring opacity and a beta-gauge mass particulate monitor, which were installed in the source stream as part of parallel research studies in progress at the time, provided supporting measurements. An instrument that was able to provide a continuous record of stack gas velocity and temperature was also used.

The instruments and sampling probes were installed in the stack of a 150-megawatt wall-fired boiler. The power generating unit was equipped with both hot and cold electrostatic precipitators containing a total of 14 stages. Therefore, the particulate loading in the stack could be varied in finite in-crements over a wide dynamic range. A small building was erected at the base of the stack to house the

Table 13.1

Instrumentation Utilized During Study

A. SO₂ Measurement
   1. DuPont 460/1 Source Monitoring System
   2. CEA/Barringer MK-IV SO₂ Stack Monitor
   3. Bailey Meter Co. SO₂ Stack Monitor

B. Supporting Measurements
   1. EPA Method 6— SO₂
   2. EPA Method 5— Mass Particulates
   3. Transmissometer— Opacity
   4. Beta Gauge — Mass Particulates
   5. PMC Autopitometer— Continuous Stack Gas
      Velocity, Temperature

C. Data Acquisition
   1. EA 2020 + Teletype
   2. Stripchart Recorders

---

instrument control units and a digital data acquisition system. The output signals from all of the monitoring devices were coupled to are Esterline-Angus 2020 digitizer with a teletype print out. In addition, each measurement was recorded on stripcharts.

Figure 13.8 is an illustration of the optics utilized in the CEA in-stack SO₂ correlation spectrometer. In this system, light from a tungsten halogen lamp is collimated and then reflected off the flat zero/read mirror. When this mirror is in the "read" position, light is directed into the probe and the probe mirror directs the light beam back into the spectrometer. If SO₂ is present in the probe slot, it will absorb energy in regularly spaced bands at 3025Å. Light passing through the entrance slit is reflected off the modulator mirror to the diffraction grating. The grating disperses the light, spatially displaying a focused absorption spectra of SO₂ at the exit mask. The optical center of the modulator mirror is tilted at a slight angle with respect to its axis of rotation, causing the angle at which light strikes the grating to vary as the motor rotates. This in turn causes the absorption spectra to scan the exit mask in a circular fashion, creating a series of harmonics whose intensity represents the SO₂ concentration.

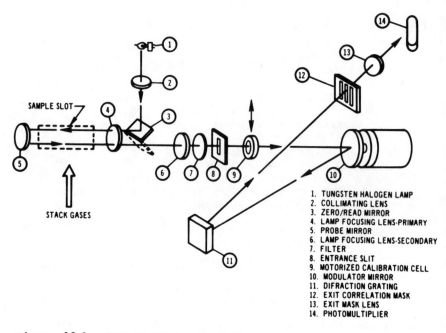

1. TUNGSTEN HALOGEN LAMP
2. COLLIMATING LENS
3. ZERO/READ MIRROR
4. LAMP FOCUSING LENS-PRIMARY
5. PROBE MIRROR
6. LAMP FOCUSING LENS-SECONDARY
7. FILTER
8. ENTRANCE SLIT
9. MOTORIZED CALIBRATION CELL
10. MODULATOR MIRROR
11. DIFRACTION GRATING
12. EXIT CORRELATION MASK
13. EXIT MASK LENS
14. PHOTOMULTIPLIER

Figure 13.8.   CEA-MK IV in-stack SO₂ monitor.

When the zero/read mirror is in the "zero" position, light is redirected into the spectrometer, by-passing the sample probe, to provide a zero check. If a temperature-stabilized gas cell containing a known SO_2 concentration is introduced into the path while the mirror is in the "zero" position, an instrument span check can be made. The output signal is affected by temperature variations of the absorbing gas in the sample slot. The effect is governed by the Charles' Law Relationship and spectral band-broadening at high temperatures. A 6°C change in stack gas temperature alters the output signal by 3%. This could be significant in applications in which there are wide fluctuations in the emission temperature
    Figure 13.9 illustrates the operation of the Bailey SO_2 source monitor. The system consists of a source housing and a receiver. The source contains two hollow cathode lamps, a reference sensor, and optics. The receiver contains another identical sensor. A slotted pipe is provided with a metered flow of purge air to each housing to maintain a definite optical pathlength through the gas to be

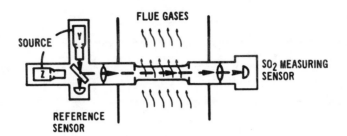

FIRST HALF CYCLE

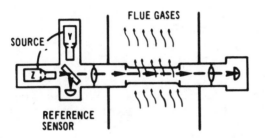

SECOND HALF CYCLE

Figure 13.9. Bailey SO₂ source analyzer.

analyzed. The electronics are contained in a cabinet.
The source lamps were chosen to emit ultraviolet
energy at two closely spaced wavelengths. In operation, the lamps are pulsed alternately on and off
and out of phase with each other. Under these conditions, the sensors are detecting the absorption of
UV energy at two discrete wavelengths closely spaced
so that the extinction coefficient of particulate
matter remains the same. Therefore, the output
signals represent the average SO_2 concentration across
the pathlength determined by the slotted pipe.
     After installation, the instruments were operated
continuously for three months. For nearly two-thirds
of this period, they were left unattended as an
attempt to assess their true reliability in an actual
installation. An example of the information being
obtained in the program is outlined in Table 13.2,
which summarizes the data from an experiment to
determine possible particulate interferences on the

Table 13.2

Effect of Increasing Stack Opacity on Instrument Response,
March 1, 1973

Sample	Opacity	Method 6	DuPont	CEA MKIV	Bailey SO$_2$
A	2.5%	460ppm	451,- 1.9%	424,-7.8%	590,+28.2%
B	3.0	475	438,- 7.8	416,-12.4	570,+20.0
C	15.0	675	577,-14.5	550,-18.5	680,+ 0.7
D	17.5	619	632,+ 2.1	599,- 3.2	720,+16.3
E	45.0	661	664,+ 0.5	624,- 5.5	730,+10.4
F	47.5	682	678,- 0.6	638,- 6.4	750,+ 9.9

measurement systems. In this experiment, the particulate concentration in the stack gas stream was
systematically increased to yield a range from 2.5%
through 48% stack opacity. During this time, the
instruments were operated continuously and the data
logger was cycled at 10-second intervals to closely
approximate intergration of each instrument signal.
Concurrently, a series of compliance test (Method 6)
samples were taken as a reference. During the course
of the experiment, net load of the boiler output
increased, which resulted in a proportional increase
in the SO$_2$ concentration by some 50% as seen from
the Method 6 analyses. Throughout the period, data
obtained from the DuPont and CEA systems did not show
an appreciable effect from the increasing particulate
level. However, the Bailey monitor did appear to
respond to the changes in opacity since the relative
error in SO$_2$ concentration as measured by the instrument was 28% higher than that of the Method 6
sample obtained at a stack opacity of 2.5%. The
relative error shows a consistent decrease with an
increasing opacity level to an error of approximately 10% at a stack opacity of 47%.

At this time, it is felt that the cause of the
interference is related to the pulse frequency of
the hollow cathode lamps. Data from the in-stack
transmissometer indicates that the particulate concentration varies dynamically over very short time
intervals. These fluctuations can occur either in
or out of phase with the pulsing rate of the Bailey

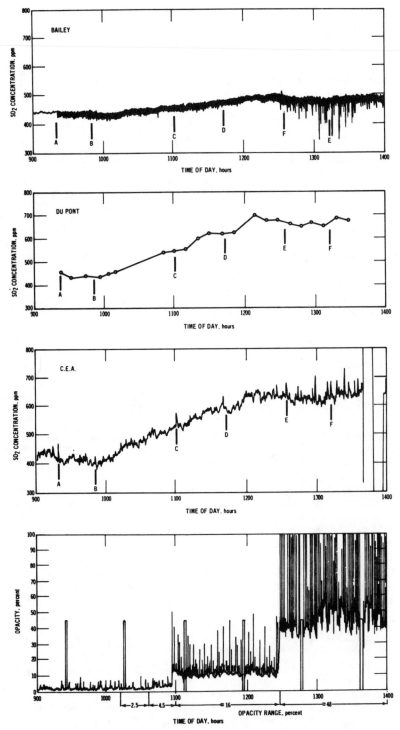

Figure 13.10. SO_2 concentration, ppm and stack opacity, %, 0900-1400, March 1, 1973.

sources. If they occur out of phase, the energy
absorbed during the pulse of one source would occur
at a higher background of particulate matter relative
to the second source, appearing as an erroneous
measurement. Figure 13.10 serves to illustrate
this characteristic behavior. In this figure, the
transmissometer recording is shown as well as repro-
ductions from the Bailey and CEA stripcharts. The
DuPont system did not contain a continuous recorder
because of its sequential operation. The analyzer's
SO_2 analysis sequence was held for the experiment
and the data logger output for the DuPont instrument
was used to represent its analysis in Figure 13.10.
At the present time, reduction of all of the monitoring
data is underway and will be presented in a future
publication. In addition, a similar study has been
planned for the fall of 1973 to investigate the
performance of certain *in situ* and extractive NO
and NO_2 monitoring systems.

SUMMARY

In summary, we have seen that there are several
approaches one can take to continuously monitor a
gaseous source emission stream. The extractive
approaches involve certain degrees of sample conditioning dependent upon not only the nature of the
emission source but also the detection technique
being employed. The conditioning techniques include
filtered probes, heated sample lines, water vapor
removal systems, or a variety of dilution devices
coupled to ambient air monitors. Dilution systems
might offer the advantage that both ambient air and
source monitoring could be accomplished by the same
instrument. A recent investigation has demonstrated
the viability of *in situ* monitoring for SO_2, which,
in effect, eliminates all sample conditioning.

REFERENCES

1. McCoy, J. "Investigation of Extractive Sampling Inter-
    face Parameters," Walden Research Corporation, EPA
    Contract 68-02-0742 (February 1973).
2. Rodes, C. E. "Variable Dilution Interface System for
    Source Pollutant Gases," *Proc., Anal. Instr., 11*, 125 (1973).
3. McKinley, J. J. "Permeation Sampling—A Technique for
    Difficult On-Stream Analyzers," *Proc., Anal. Instr., 10*,
    214 (1972).

4. Rodes, C. E., R. M. Felder, and J. K. Ferrell. "Permeation of Sulfur Dioxide Through Polymeric Stack Sampling Interfaces," *Environ. Sci. Technol., 7,* 545 (1973).
5. Homolya, J. B. and R. J. Griffin. "Abstracts," 164th National Meeting of the American Chemical Society, New York, New York, No. WATR-84 (August 1972).
6. Hedley, W. H. "Construction and Field Testing of a Commercial Prototype Disc Diluter," Monsanto Research Corporation, EPA Contract 68-02-0716 (January 1973).
7. *Federal Register*, "Method 6—Determination of Sulfur Dioxide From Stationary Sources,"" Part II, *36*(247), 24890-24891 (1971).

CHAPTER 14

APPLICATION OF A BETA GAUGE TO THE MEASUREMENT OF
MASS EMISSIONS IN A COAL-FIRED POWER PLANT

John W. Davis

Several techniques are under development at EPA
for continuous mass effluent monitoring of particulate
matter. One of the most promising is mass determination by absorption of beta radiation. Instruments
effecting this principle have been commercially available for some time but have not come into general use.
There is, however, an increasing need for an automatic
monitoring system that can be used with confidence
on a variety of sources.
    In this series of tests, instrumentation was
installed in a coal-fired power plant. A commercially
available beta gauge was operated and a series of
manual samples were taken. The results indicate that
mass gravimetric determinations from the manual
sampling train compared favorably with the beta-gauge
readings.

INTRODUCTION

    One area of current interest at the Environmental
Protection Agency is the development of a continuous
mass monitor for stationary sources. The current
manual methods for source sampling are cumbersome and
time-consuming. When an industrial process is variable,
the manual sampling results may be questioned since
they are often obtained only during a part of the
operational cycle. An on-line instrument taking
samples continuously over the whole cycle of an

operation will obviously provide better and more complete information.

After a review of the state-of-the-art in particulate sampling sponsored by EPA,[1] a program was initiated to build an instrument with beta-ray attenuation as a method of determining the mass collected on a filter media. This method was chosen because it is, in principle, independent of particle size and chemical composition of the sample. Earlier research conducted by Nader and Allen,[2] Dresia,[3] and Bulba and Silverman[4] confirmed that the beta attenuation method gave results that compared favorably with mass determinations obtained gravimetrically.

Two instruments were built for EPA under contract[5] and tested on a power plant stack. The hot, wet corrosive stack gas had to be diluted prior to being introduced into the instrument. A system providing three parts filtered dry air to one part stack gas was used. At this dilution rate, both instruments sustained some internal damage. In addition, one of the units had an electronic failure that was attributed to the severe atmosphere in the stack environs. From these tests, it became obvious that the electronic circuitry had to be protected and that a built-in stack interface system was necessary for easy instrument usage.

A German instrument recently introduced on the American market seems to have solved the major problems. The instrument is a ruggedly built, fully integrated system that maintains the stack gas at a high temperature through the entire sampling unit, thus obviating the need for any dilution system. An instrument was acquired from the American distributor and series of field tests, designed to compare the mass determinations measured by the beta gauge with those obtained from the manual sampling method, were initiated. The purpose of this work and the work to follow is to develop a set of specifications for continuous or semicontinuous monitors of particulate emissions from stationary sources. EPA's use of this particular device should not be construed as an endorsement.

THE BETA GAUGE DESCRIPTION
AND OPERATION

Figure 14.1 is a diagram of the instrument. The stack gas enters through a stainless steel nozzle and probe; a ball valve separates the probe from the

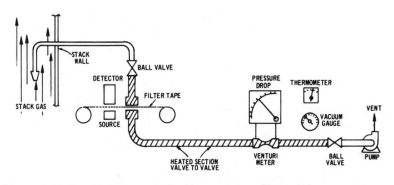

Figure 14.1. Beta gauge instrument configuration.

instrument. This valve is opened and closed automatically and serves to isolate the instrument from
the stack when no sample is being drawn through the
filter tape. Both the probe and the valve entrance
and exit are wrapped with heating tape. The stainless steel tube running from the first ball valve
to the second is also heated. The velocity of the
stack gas entering the nozzle is controlled by the
venturi meter operating in conjunction with the second
ball valve. A pair of points are set on the meter
to maintain a constant pressure drop across the
venturi. These points control the second ball valve.
As the meter needle touches the upper or lower point,
the ball valve automatically opens or closes to
admit a sufficient amount of outside air. A constant
flowrate is thereby maintained by the vacuum pump.
     During sampling, the glass fiber filter media
in tape form is clamped by a bellows arrangement.
Plastic strips along the outside edges of the tape
give it sufficient strength to prevent tearing as
it moves through the sample cycle. If the tape
should tear, a tape monitoring device shuts the
instrument down. The beta source is carbon 14 and
the detector is a Geiger-Muller tube. A computation
section and recorder display complete the system.
     Before sampling can begin, the velocity of the
stack gas at the sample point must be determined by
standard methods. On the basis of this information,
a nozzle of appropriate inside diameter can be
selected from a graph provided with the instrument.
One of the major considerations in particulate
sampling is the maintenance of isokinetic conditions.
This means that the velocity in the nozzle must equal
the velocity of the stack gas at the sampling point.

In this instrument isokinetic sampling is approximated
by selecting the proper nozzle and by setting the
correct pressure drop at the venturi meter as determined through a series of test runs and calculations.
As indicated above, the flowrate is held constant
by the set points on the meter. The instrument
provides for the selection of different sampling
times. Once a time period is set, a constant volume
of gas is pulled through the tape for each sample.
A temperature of approximately 150°C is maintained
from the stack wall to the pump, thus preventing
condensation in the system.

The measurement cycle begins with one minute
background count taken on the filter. The spot is
indexed into the sample chamber and a sample is
pulled through the tape for the preset time period.
The spot is then moved back under the detector and
the second count is made. The computation in
milligrams is displayed on a recorder as a 20-second
spike. Since the sample volume is held constant,
the only variable is the amount of deposit. A
comprehensive description of the instrument and the
beta absorption technique is given by Dresia and
Spohr.[7]

THE MANUAL SAMPLING TRAIN

The mass concentrations recorded by the beta
dustmeter were compared with the results obtained
by an EPA manual sampling train (Figure 14.2). The
stack gas enters through a buttonhook nozzle and is
pulled through a heated tube into the sample box
housing. In both the probe and the sample box, the
temperature is maintained at 120°C. Although a
prefiltering cyclone as shown in the figure is used
in some cases, one was not used in these tests.
Following the filter is a set of impingers in series.
These impingers are set in an ice bath; the first
two contain 100 ml of water, the third is dry and
the fourth contains 200 g of silica gel. An umbilical
cord carrying the electrical connections, the pitot
tube lines, and the sample line connects the sample
box to the control box. The control unit houses
valves, a dry gas meter, a manometer, and the vacuum
pump.

Built into the probe assembly is an S-type pitot
tube and a thermocouple (not shown on this sketch).
The thermocouple has a separate direct readout.

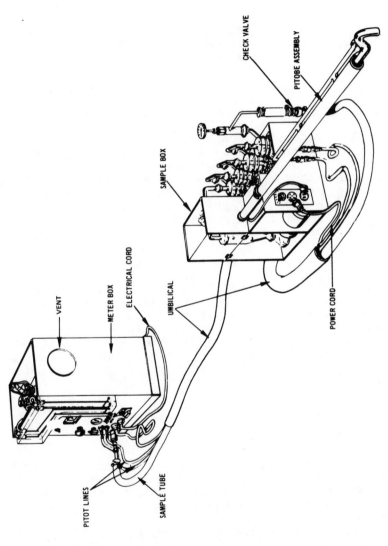

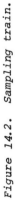

Figure 14.2. Sampling train.

Isokinetic flow is achieved and maintained by setting a second manometer with a valve that controls the vacuum pump. The set point on the manometer is obtained from the measurement of various parameters and the use of a nomograph. A dry gas meter records the sample volume. For a complete description of the train and its operation, see the paper by Smith, *et al.*[8] The Federal Register[9] contains the standard procedures to be followed for manual sampling.

THE FIELD SITE

The field site was a coal-fired power plant near Charlotte, North Carolina. The stack was three meters in diameter and had a one-centimeter thick steel wall with no liner. A hot electrostatic precipitator was used to control emissions. The duct and stack configuration made it impossible to locate the sampling site the specified eight stack diameters downstream from any disturbance. In this case, the sample site was approximately three diameters from the nearest bend. The velocity profile, therefore, varied considerably across the stack, but did not vary with time. This was illustrated by the observation that when one returned to the stack after a week or a month's time, the profile could be predicted from previous measurements but the velocity varied from 30 to 100 feet per second in the stack itself. The average velocity in the stack was about 70 feet per second and the average temperature in the stack was about 150°C.

EXPERIMENTAL PROCEDURE

The standard operating procedure for the manual train was followed in so far as it served the purpose of the experiment. Deviations from the procedure other than the one noted in the preceding section are described below.

Two different sampling techniques with the manual train were used. In one case the samples were taken by traversing the stack in the traditional way (Figure 14.3). The stack area for these tests was divided into four equal areas and five-minute samples were taken at the centroid of each area. The probe was admitted to the stack through two ports located

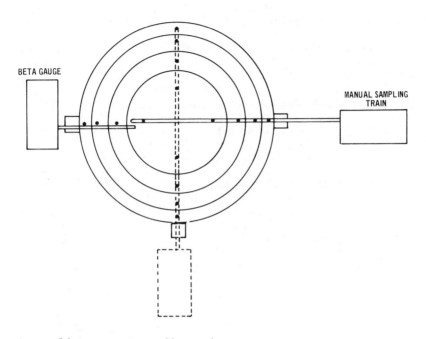

Figure 14.3. Stack configuration.

at 90° from each other. Samples were taken at each
point and the flow adjusted to achieve isokinetic
sampling. Since there were 16 points, the total
sample time was 80 minutes.

The second sampling technique was to take a
sample at a single point. The probe was set along-
side the beta gauge probe and a one-hour sample was
taken. The beta gauge probe and the manual sample
probe were located at a point in the stack with
average velocity.

The maximum sampling time on the beta gauge as
received from the manufacturer was six minutes. In
most cases the selected sampling time was 200 seconds.
When the loading went up, the sampling time had to be
cut back to 100 or 50 seconds and in one case, 20
seconds. The time between the start of one sample
and the beginning of the next ranged from about 3 to
6 minutes. The additional time was needed to index
and read the tape. This meant that the total number
of beta gauge samples taken simultaneously with a
manual sample ranged from 10 to 20.

Prior to the start of the measurement cycle, the
zero and calibration check built into the beta gauge

instrument were conducted. Next, a series of manual
samples were taken by either the traverse or single
point technique over an eight- to ten-hour period.
A zero check was taken at the end of the day. As
the manual sampling train was being prepared, calibration checks of the beta gauge were made periodically
by inserting some preweighed pieces of sample tape
into the beta gauge for a sample cycle. The filter
was reweighed after exposure and the weight of the
material on the filter was compared with the beta
gauge reading for a direct calibration.

    When following the EPA method of determining the
concentration of particulate matter in the stack gas,
one must include the rinse from the nozzle, probe,
and filter holder as well as the filter itself.
This is traditionally called the front half of the
catch. When the particle loading in the impingers
that follow the filter is included, the sum is known
as the total catch.

ANALYSIS OF RESULTS

    The particle matter collected on the filter in
the EPA train (determined gravitrically) is compared
with the average beta gauge reading over the same
time period (Figure 14.4). The line is a least
squares fit with a slope of 1.09 and an intercept
of -7.82. These data include all the traverse
points and all the single point samples taken over
a several week period.

    When the samples acquired by using the traverse
method were compared with the single point information, no statistical differences could be found,
and therefore, the data were combined in the analysis.
A total of 31 paired points are shown in the figure.
One interesting aspect should be noted: all of
these tests were run while the precipitator was
operating at full efficiency and yet the particle
load in the stack varied from about 10 to 70 mg/m^3.
It is obvious that if only three samples are taken
(as is recommended in some procedures) the results
could be misinterpreted.

    Figure 14.5 shows the front-half catch in the
manual train plotted with the beta gauge readings.
The slope of the line is 0.46 and the intercept is
-11.00. In this case, only 14 points were available
for comparison since the particle matter deposited
in the nozzle, probe, and filter holder was measured
only when the standard traverse method was used.

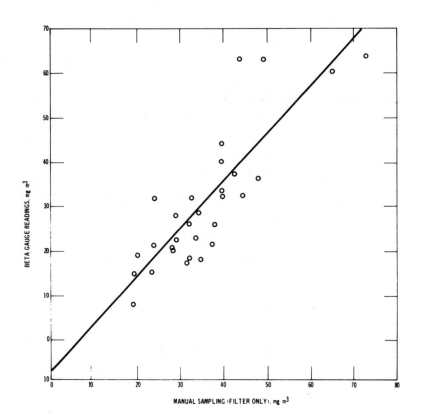

Figure 14.4.   Manual sampling (filter only) vs. beta gauge.

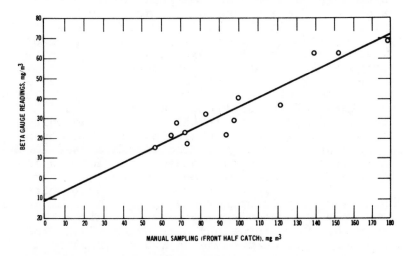

Figure 14.5.   Manual sampling (front half catch) vs. beta gauge.

Method 5 as described in the Federal Register[9] calls
for using this front-half catch to determine compliance
with regulations. The least squares fit line indicates
that one would have to multiply the beta gauge reading by about 2-1/2 to 3 in order to get the front-half
catch information.
     A total of 39 manual samples were taken of which
35 were good. Three of the traverse type samples
and one of the single point samples were lost for
various reasons. Four other data points are not included in this analysis because they were taken when
the electrostatic precipitator was cut back.
     Table 14.1 summarizes the results. Excellent
agreement between the filter catch only and the beta
gauge reading was obtained. As indicated above, the

Table 14.1

Summary of Results

	No. of Samples	Mean	Standard Deviation	Correlation Coefficient
Filter only	31	36.6	13.0	R = 0.86
Beta gauge readings	31	32.0	16.3	R = 0.94
Total catch	14	96.5	33.8	

relationship between the front-half catch and the
beta gauge is less encouraging. This may be due to
particle losses in the nozzle and probe and at the
right angle bends. Examination of these parts did
show some deposits. The major losses, however,
occurred at the expansion joint down stream of the
ball valve. This area should have been wrapped with
heating tape, but was not during these tests. The
manufacturer of the instrument maintains that the
losses would have been markedly reduced had this
been done. If losses at the ball valve had not been
sustained, the results would have been closer to the
front-half catch results. Whether the large differences would have been eliminated cannot be ascertained
from these data. There is good statistical correlation between the front-half catch and the beta gauge,
which would indicate that once the relationship
between the two methods had been established, the
beta gauge could be used as an indicator of total

emissions. This would probably require calibration
at each individual installation and require that no
major changes take place in the process or control
equipment.

Figure 14.6 indicates the practical use to which
the beta gauge may be put. The clock time is plotted
against the beta gauge reading. During this particular day, the electrostatic precipitators were cut

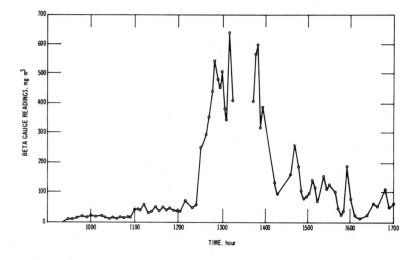

Figure 14.6. Variation of particle emissions with time.

back in stages beginning about 10:45 a.m. and ending
about 2:00 p.m. The quick response and semicontinuous
reading of the instrument could be used to evaluate
the performance of control devices, and the instrument can obviously be used as an on-line source
monitor.

These were preliminary tests, and more are
scheduled for the future on various types of stationary sources. A number of problems were experienced
in using the instrument. Perhaps the most critical
was the 50-60 cycle problem. The instrument as received from Germany operated at 50 cycles; therefore,
all clock-type mechanisms did not operate correctly.
This caused the programmer to malfunction on occasion
so that the instrument needed constant attention.
It is probable that the manufacturer will correct
this condition prior to delivering any other instruments

in this country. The only item on the instrument
that seems to need to be changed is the sample time.
This would also affect the source and the detector.
Under the present configuration, the maximum sample
time is six minutes. When the concentration goes up
(as it did when the precipitators were cut back)
the sample time must be reduced to extremely short
intervals. In the case shown in Figure 14.6, the
sample time for the heavy concentration had to be
set at 20 seconds to keep the reading on scale.
Ultimately a variable scale should be built into
the instrument and sample times up to one hour should
be possible.

SUMMARY AND CONCLUSIONS

A series of field tests **was** conducted in which
mass determinations obtained by the beta attenuation
method were compared with manual sampling results.
The beta gauge results were compared with the filter
of the manual sampling train and with the front half
of the catch. The use of the beta gauge as a semicontinuous monitor was also demonstrated and
discussed.
This work indicates that the beta absorption
technique can be an effective way to measure particle
concentration. The excellent correlation between
the beta attenuation method and the gravimetric
method is encouraging. The major problem is the
losses in the lines between the stack and the
measuring device. Further tests must be conducted
to determine the severity of this problem.
Additional field information is also needed to
ascertain the applicability of the instrument to a
variety of source types.

REFERENCES

1. Sem, G., J. Borgos, K. Whitby, and B. Liu. "State-of-
   the-Art: 1971, Instrumentation for Measurement of
   Particulate Emission from Combustion Sources," Final
   Report (4 volumes) for EPA on Contract No. CPA 70-23
   (1971-1973).
2. Nader, J. S., and D. R. Allen. "A Mass Loading and
   Radioactivity Analyzer for Atmospheric Particulates,"
   Am. Ind. Hyg. Assoc. J., 21, 300 (1960).

3. Dresia, H. "Continuous Measurement of the Dust Content in Air and Exhaust Gases by Radiation," (in German) *VDI-Z, 106,* 24 (1964).
4. Bulba, E., and L. Silverman. "A Mass Recording Stack and Monitoring System for Particulates," APCA Annual Meeting, Paper No. 65-141, Toronto, Canada (1965).
5. Lilienfeld, P., and J. Dulchinos. "Vehicle Particulate Exhaust Mass Monitor," Final Report for EPA on Contract No. 68-02-0209 (1972).
6. Duke, C. R., and B. Y. Cho. "Development of a Nucleonic Particulate Emission Gauge," Final Report for EPA on Contract No. 68-02-0210 (1972).
7. Dresia, H., and F. Sphor. "Experience with the Radiometric Dust Measuring Unit," *Staub-Reinhalf Luft* (English translation) *31*, 6 (1971).
8. Smith, W. S., R. M. Martin, D. E. Durst, R. G. Hyland, T. J. Logan, and C. B. Hager. "Stack Gas Sampling Improved and Simplified with New Equipment," APCA Annual Meeting, Paper No. 67-119, Cleveland, Ohio (1967).
9. *Federal Register*, Part II, *36(247),* 1971.

INDEX

INDEX

301

ANALYTICAL METHODS APPLIED TO AIR POLLUTION MEASUREMENTS

ROBERT K. STEVENS, Research Chemist, and WILLIAM F. HERGET, Research Physicist, Environmental Protection Agency, Editors

This book represents a major step toward improved air pollution measurement of undesirable atmospheric components. For the first time—in many cases—it brings together the latest technological advances in this field by chemists, physicists and electrical engineers, including the application of lasers and other electro-optical devices to measure pollutant concentrations.

Analytical Methods Applied to Air Pollution Measurements is organized in three main sections. Section one concerns methods to measure gaseous pollutants at ambient concentrations; Section two discusses procedures that characterize the properties of particulates in the atmosphere; and Section three describes instrumental methods for analyzing gaseous and particulate pollutants at source concentrations.

Broadly **and** specifically this volume covers gas-filter correlation spectroscopy, comparison of classical and newly developed x-ray fluorescent methods for measuring elemental composition of particles, and advances in techniques for extracting gaseous and particulate pollutants from power plant stacks. Further chapters describe the application of chemiluminescent techniques to measure ambient concentrations of gaseous pollutants, and descriptions of two new techniques to measure sulfuric acid vapor at ambient temperatures. These reports are profusely illustrated with 136 figures and 29 tables.

Diverse and comprehensive, this book will be valuable to anyone who needs to know the most recent research findings in air pollution measurement and wants to have the complete picture. Its use will be widespread among analytical chemists, environmentalists, state and local air pollution agencies, air pollution control officials in the chemical, power-producing industries and manufacturers of air pollution measurement instrumentation.